AF847058

Springer Proceedings in Energy

The series Springer Proceedings in Energy covers a broad range of multidisciplinary subjects in those research fields closely related to present and future forms of energy as a resource for human societies. Typically based on material presented at conferences, workshops and similar scientific meetings, volumes published in this series will constitute comprehensive state-of-the-art references on energy-related science and technology studies. The subjects of these conferences will fall typically within these broad categories:

- Energy Efficiency
- Fossil Fuels
- Nuclear Energy
- Policy, Economics, Management & Transport
- Renewable and Green Energy
- Systems, Storage and Harvesting
- Materials for Energy

eBook Volumes in the Springer Proceedings in Energy will be available online in the world's most extensive eBook collection, as part of the Springer Energy eBook Collection. Please send your proposals/inquiry to Dr. Loyola DSilva, Senior Publishing Editor, Springer (loyola.dsilva@springer.com)

More information about this series at http://www.springer.com/series/13370

V. Kamaraj · Jayashri Ravishankar ·
S. Jeevananthan
Editors

Emerging Solutions for e-Mobility and Smart Grids

Select Proceedings of ICRES 2020

Springer

Editors
V. Kamaraj
Sri Sivasubramaniya Nadar (SSN) College
Kalavakkam, India

S. Jeevananthan
Pondicherry Engineering College
Pondicherry, India

Jayashri Ravishankar
School of Electrical Engineering
and Telecommunications
New South Wales, NSW, Australia

ISSN 2352-2534 ISSN 2352-2542 (electronic)
Springer Proceedings in Energy
ISBN 978-981-16-0718-9 ISBN 978-981-16-0719-6 (eBook)
https://doi.org/10.1007/978-981-16-0719-6

© The Editor(s) (if applicable) and The Author(s), under exclusive license to Springer Nature Singapore Pte Ltd. 2021
This work is subject to copyright. All rights are solely and exclusively licensed by the Publisher, whether the whole or part of the material is concerned, specifically the rights of translation, reprinting, reuse of illustrations, recitation, broadcasting, reproduction on microfilms or in any other physical way, and transmission or information storage and retrieval, electronic adaptation, computer software, or by similar or dissimilar methodology now known or hereafter developed.
The use of general descriptive names, registered names, trademarks, service marks, etc. in this publication does not imply, even in the absence of a specific statement, that such names are exempt from the relevant protective laws and regulations and therefore free for general use.
The publisher, the authors and the editors are safe to assume that the advice and information in this book are believed to be true and accurate at the date of publication. Neither the publisher nor the authors or the editors give a warranty, expressed or implied, with respect to the material contained herein or for any errors or omissions that may have been made. The publisher remains neutral with regard to jurisdictional claims in published maps and institutional affiliations.

This Springer imprint is published by the registered company Springer Nature Singapore Pte Ltd.
The registered company address is: 152 Beach Road, #21-01/04 Gateway East, Singapore 189721, Singapore

Preface

Volume 2: e-Mobility and Smart Grids

The escalating importance and concern for building a sustainable society in future emphasizes the need for environment safe systems and inventions. Several alternatives like e-mobility and usage of renewable sources are gaining paramount significance in this era. In contrast to the conventional grid-based systems, the digitalized world is in need of smart meters and appliances that can ease the pressure on humans. This book sheds light on these versatile technologies that are finding great popularity and usage. The major topics covered in this book include electrical machines for electric vehicle, power converter design, energy storage systems and smart grid.

This book is a compilation of full-length papers presented during the proceedings of the First Virtual International conference on Renewable Energy Systems (ICRES 2020) held during August 26–28, 2020 organized by department of Electrical and Electronics Engineering, Sri Sivasubramaniya Nadar College of Engineering, Kalavakkam. The conference was an eye-opener for the academic, industrial as well as the research community in bringing out the various ideas and innovations in this domain of e-mobility and smart grid systems.

The conference attracted international as well as national participation from many crème de la crème institutions that includes paper presentations from IITs, NITs, IITDM and Anna University. This three-day event had five keynote speakers who delivered top-notch lectures on various topics including robotics, photovoltaic systems, smart grid applications etc. The papers that have been included in this book have been reviewed by the domain experts to ensure the quality and correctness of the manuscript. ICRES 2020 technical committee and editors thank the authors for their support during the review process. This book will serve as a cradle of innovative ideas and applications with respect to this domain which can help in building a green and clean future.

Kalavakkam, India V. Kamaraj
Sydney, Australia Jayashri Ravishankar
Puducherry, India S. Jeevananthan
January 2021

Contents

About the Editors

Dr. V. Kamaraj Professor and Head in the Department of Electrical and Electronics Engineering, SSN College of Engineering, Chennai, has 30 years of teaching and research experience. He has received his B.Tech. (EEE) degree first class from Calicut University in 1987, M.E. in Power Electronics and Ph.D. in Networks from Anna University, respectively, in the year 1993 and 2001. He has published over 100 research publications in refereed international and national journals and conferences. He served in Sri Venkateswara College of Engineering from 1989 to 2006 and was elevated from associate lecturer to various positions. After joining SSN College of Engineering in 2006 as a Professor in the Department of EEE he was instrumental in getting MODROB project and seminar grant from AICTE. He has organised international conferences and several workshops at SSN College of Engineering. Presently he is guiding 5 Ph.D. research scholars and 12 scholars have completed Ph.D. under his guidance. He is currently working in the area of power quality, machine design and power converter topologies. He has received ISTE Periyar Award, Best Engineering College Teacher, and IET Diamond Salute Award in 2014. He is the principal investigator for the project titled "Design and Development of Permanent Magnet Assisted Reluctance Motor Drives for Pump Application" funded by Science and Engineering Research Board (SERB). He is an active member in IEEE and life member in ISTE.

Dr. Jayashri Ravishankar received her BE, ME and Ph.D. all in Electrical Engineering from Anna University, India. After 15 years of teaching in India, she joined the University of New South Wales (UNSW), Sydney in 2010 as an API (Australian Power Institute) sponsored academic for promoting power engineering education in Australia. Her teaching and research interests include power system modelling, analysis and control, renewable energy integration, smart grids and micro grids. She has authored and coauthored more than 80 journal and conference papers in this area. Jayashri set up the first microgrid test bench at UNSW in 2012 and has since supervised to completion seven Ph.D. students in the area of microgrids. She is the associate editor of the *Australian Journal for Electrical and Electronics Engineering*and a regular reviewer for several IEEE, IET and other journals and conferences. She has research active collaborations in India, Ireland and USA. Jayashri also has a deep interest in learning and teaching, and consistently implements strategies using technology innovations and industry partnerships to improve students' active learning. She is institutionally and internationally recognized for the impact of her innovative, research-led and highly effective teaching and leadership. Alongside high-quality research outcomes and international collaborations, she leads best-practice advanced teaching in electrical engineering through imaginative initiatives, including blended industry lectures and flipped mode strategies. By demonstrating teaching excellence in her professional, inclusive learning environments, particularly supportive of international students, she has been very influential in educational peer leadership. By mentoring her students to focus on learning as future professionals, and by excelling in UNSW leadership and service roles, she shares her passion for engineering education and research. Esteem indicators include Senior Fellowship of the Higher Education Academy awarded by Advance HE (UK), and individual Teaching Excellence Awards both from the Faculty of Engineering and the UNSW.

Dr. S. Jeevananthan is Professor in the Department of Electrical and Electronics Engineering, Pondicherry Engineering College, Puducherry. He has 20 years of teaching and research experience. He has received his B.E. (EEE) degree from Madurai Kamaraj University in 1998, M.E. in Power Electronics and Drives from Bharathiyar University in 2000 and Ph.D. in Power Electronics from Pondicherry University in the year 2008. He has published over 150 research publications in refereed international and national journals and conferences. He has published 3 books and has completed 2 funded projects. Presently he is guiding 4 Ph.D. research scholars and 6 scholars have completed Ph.D. under his guidance. He is currently working in the area of power converter topologies, control strategies and AC drives. He has completed consultancy projects for Numeric power systems Ltd. and Consul consolidated Pvt. Ltd. He visited Singapore, Malaysia, Srilanka, Germany, France and Netherland. He is currently board of studies member in several institutions.

Chapter 1
Intelligent Stand-alone Solar PV-Enabled Rainwater Harvester and Power Generator

R. S. Sibi, T. V. Sharath, A. Khaja Najumudeen, and R. Rajeshkanna

Abstract According to UNESCO, with the existing climate change scenario, by 2030, water scarcity in some arid and semiarid places will displace between 24 million and 700 million people. In India, the paucity of water is increasing every year in a drastic way and so does the demand. This has occurred as a result of deficient planning and inappropriate management of water resources. Surface water and groundwater, which constitute approximately 40% of total precipitation, are two major sources of drinking water. It is evident that the collection of rainwater is very pivotal and beneficial. This paper illustrates the idea of conserving rainwater in an economic and renewable way by occupying free space and compact arrangement. This stand-alone system also focuses on automatic functioning of system using IoT.

Keywords Automation · IoT · Rainwater harvesting · Solar PV · Solar tracking · Stand-alone

1.1 Introduction

The paramount sources of all drinking water and freshwater are considered to be snowmelt and rainwater in this world. By 2025, approximately 1.8 billion people are predicted to reside in places with extreme water paucity [1]. To subsist with inter-annular variation in the precipitation and maintain well-being of human, rainwater harvesting strategy has long been carried out around the world [2]. At present, the loss of water is estimated to be 72.5%, which could be utilized for various essential household applications [3]. Smallholder agriculture is largely dependent on irrigation contributed by many small rainwater harvesting reservoirs known as tanks, in South Indian State of Tamil Nadu [4].

Solar-enabled rainwater collector is a device used for collection of rainwater abruptly by ensuring that it avoids runoff into water bodies or does not immerse into the ground or does not get contaminated. The major restriction of this option is the

R. S. Sibi (✉) · T. V. Sharath · A. Khaja Najumudeen · R. Rajeshkanna
Hindusthan College of Engineering and Technology, Coimbatore, Tamil Nadu, India

© The Author(s), under exclusive license to Springer Nature Singapore Pte Ltd. 2021

V. Kamaraj et al. (eds.), *Emerging Solutions for e-Mobility and Smart Grids*,
Springer Proceedings in Energy, https://doi.org/10.1007/978-981-16-0719-6_1

absence of rainwater in the year [5]. But this option can be utilized as alternative sources of water if rainwater is optimally stored during the rainy season [6].

Optionally certain filters are applied, through which useful and good-quality water can be obtained. Allowing an increase in economic productivity, rainwater harvesting systems enable farmers to overcome drought and to expand farming potentials throughout the year [7, 8]. Access to greater economic profits can eventually minimize the poverty and optimize the quality of life for rural society [9].

Technical design and identification of suitable sites are the major factors at which rainwater harvesting system is highly dependent [10]. For appropriate and efficient rainwater harvesting system, various methodologies have been emerged for selection of sites and techniques [11]. For selecting suitable sites and rainwater harvesting techniques in small areas, field surveys are commonly adopted [12]. Selection of suitable sites for various rainwater harvesting technologies in larger areas is a crucial problem [13].

1.2 Study of Existing System

The rooftop rainwater collection is considered as an additional source of potable water in developing countries. Generally, a rooftop rainwater harvesting system comprises a rain collection roof area, a water storage system, a treatment system, a supply system and piping.

Components such as filters, distribution pipe work, pumps, overflow and backflow prevention devices are additionally utilized for various purposes. Connection of a gutter system to the end of roof enables the rainwater that runs off from roof to be collected and stored in a tank.

The rooftop rainwater system is a commonly used method for reduction of the demand on waste treatment facilities and public water supplies. The rainwater is applicable for a wide range of applications in which treatment is not required such as clothes washing, garden irrigation and flushing.

Domestic rain harvesting systems are intended to optimize the usage of rainwater within houses and for land plots. The rain harvesting systems also invoke various challenges. One major concern for domestic rain harvesting is contamination due to chemical components present on the roof surface. Another major concern is enabling breeding grounds for insects when water is left in the gutter. This can be resolved by capturing maximum amount of water and transmitting the water from gutter to the storage tank swiftly with minimum losses. In places which do not have access to necessary municipal water soon after the monsoon months, collection of rooftop water, after proper filtration, would be highly beneficial. Another way of harvesting the rainwater is the driveway runoff harvesting. The driveway area is considerably quite larger than rooftop area in the case of abounding number of houses and office complexes. Rainwater constituted in this area would be copious, and in inclusion to that, adequate amount of rooftop water will also be added to this, which would flow through the gate. This rainwater is produced to recharge well by obstructing it with the help of a bump near the gate or a shallow gutter. As the runoff might consist of large quantity of silt, the driveway runoff should not be directed to a source well.

1.2.1 Importance of Rainwater Collector System

The rainwater collection system is mainly employed for two important aspects. The first aspect is the collection of rainwater for instant usage, and the second aspect is recharge into the soil by natural means or artificial means for sustainable and optimized groundwater source. Places like Chennai (India) were highly dependent on groundwater source, whose quality and quantity were remarkable until few years back. The open space got reduced due to urbanization, which obstructs the natural recharge and overexploited the groundwater. For approximately two months in every 2 to 3 years, over 20 lakh people strive for freshwater due to a swiftly increasing urban population in Chennai. The sources of water in Chennai highly rely upon rainfall. Puzhal, Malayambakkam, Poondiand and Cholavaram, the four prominent city reservoirs, which constitute a total capacity of 11,500 million cubic feet, have dropped to its all-time low value in 2019. One of the largest reservoirs in Tamil Nadu, Sholayar, got entirely dried up.

During monsoon season, urban areas commonly witness the problem of flood and during other seasons, scarcity of freshwater. Both these constrains arise predominantly due to the insufficiency of rainwater harvesting in such areas.

The groundwater division of the Public Works Department states that, from May 2018 to May 2019, the average increase in water level of 3.37 m below ground level was witnessed in Coimbatore (India). The maximum increase in water level (about 22.87 to 10.55 m below ground level) was observed in Sarkar Samakulam Block, in Coimbatore.

1.3 Design of Stand-alone Rainwater Harvester System

The designing process of the stand-alone rainwater collector system includes the controlling of various parameters and operation of the system components (Fig. 1.1).

1. Solar PV module

The solar PV module is used to convert the energy from solar radiation to electrical energy. The solar PV module acts as the main source of electrical energy in this system. The function of solar panel in this system is to store energy and provide supply to the system components.

The solar PV panel employed in this system has the capability of delivering output power of 100 W. The solar panel has a height of 3.3 ft and width of 2.1 ft.

Fig. 1.1 Real-time model of intelligent stand-alone solar PV-enabled rainwater collector and power generator

$$P = W_P * H * 75\% \tag{1.1}$$

where

P Daily watt-hours
W_P Solar panel watts
H Average hours of sunlight

The solar panel is installed with a tracking system to track the solar radiations. The solar tracker is adopted in order to obtain the maximum output power from the system. The solar tracking system can be graded into two groups: single-axis and dual-axis tracking system. A single-axis tracking system enables the solar panel to move along one axis, whereas the dual-axis tracking system enables the solar panel to move along two axes.

This system employs single-axis tracking system. This tracking method consists of two light-dependent resistors (LDRs) and one-geared DC motors with control unit. Here, Raspberry Pi is used as the controller to control the operations of the geared DC motors. Figure 1.4 illustrates the tracking system operation.

2. Controller system

The solar-enabled rainwater collector and power generator system consist of the following devices for controlling the various parameters in the system.

- Raspberry Pi
- Voltage controller

- Charge controller

The Raspberry Pi employed in this system is used to perform the motor controlling operations and to display the result. The Raspberry Pi model utilized here is Raspberry Pi 3 Model B. The Raspberry Pi 3 Model B comprises 1 GB RAM and BCM43438 wireless LAN.

The voltage regulator is used to regulate the power supply to the Raspberry Pi. The voltage regulator converts the 12 V supply from the battery to 5 V DC supply compatible for the Raspberry Pi. The IC used for voltage regulation is LM7805. Some amount of energy gets dissipated in **IC 7805 voltage regulator** in the form of heat. The heat is generated due to difference in the input voltage and output voltage values. The heat generated or dissipated will be more if the difference between the input voltage and output voltage is high.

$$W = (V_i - 5) * I_o \tag{1.2}$$

where

W Heat generated
V_i Input voltage
I_o Output current

An maximum power point tracking (MPPT) charge controller is adopted in this system as a charge controller. The MPPT charge controller is connected in between the solar panel and the battery. The MPPT controller is used to optimize the match between the solar panel module and the battery.

3. Rain detector

The rainwater sensor module employed in the system is connected to Raspberry Pi. The sensor module operates at 5 V DC supply. On detection of the presence of the rainwater, the signals from sensor are delivered to Raspberry Pi for further control operations.

4. Motors

The stand-alone rainwater collector system makes use of two motors as given below.

- DC series wound motor
- DC geared motor

The DC series wound motor acts as the main motor and is connected to the shaft which is used to open and close the sheets. The DC series wound motor has the capability to deliver power output of 320 W and has operating voltage of 12 V. A gear system is attached to this motor in order to reduce the speed and increase the torque.

The DC geared motor is employed in the solar tracking system. This motor is used to rotate solar panel in a particular angle depending upon the solar radiation detected by LDRs.

5. Energy storage system

A rechargeable DC battery is utilized to store electrical energy generated by the solar module. A lead–acid battery which has the capacity to store 35 Ah is employed in this system. The battery has the operating voltage of 12 V.

The battery provides the electrical energy supply to motors and Raspberry Pi. It is also used to provide supply to external circuits.

6. Water storage system

The water storage system for a stand-alone rainwater collector system can be done in two techniques: One technique is to attach a water storage tank underground, and the other technique is to attach a tank externally to store the rainwater. A tank which is capable of collecting 38 L of the rainwater is attached to this system. The tank has dimension of $0.5 \times 0.24 \times 0.32$ (in meters).

$$L = V_{\mathrm{T}} * 1000 \tag{1.3}$$

where

L Water inside tank in liters
V_{T} Volume of the tank

From Eq. 1.3, the tank has the capacity to collect approximately 38 L of rainwater.

In order to filter the incoming rainwater, a filtration system is employed to eliminate the impurities available in the incoming rainwater. In this case, the application of the carbon filtration system is applied.

7. Rain collector system

In order to collect fresh rainwater, it is extremely important to have a collecting surface. The utilization of tightly woven nylon fabric sheet is done as a collecting surface in this system. The nylon fabric sheet employed is coated with polyurethane (PU), in order to obtain its ability to become waterproof.

The sheet can cover total area of 1.2676 m^2 for collecting the rainwater. The maximum amount of rainwater collected by the collecting surface can be determined by the following equation.

$$R = A * (\text{rainfall} - B) * A_{\mathrm{S}} \tag{1.4}$$

where

R Total rainwater collected in liters
A_{S} Area of surface
A Runoff coefficient of collection which is around 80%
B Losses that occur due to wetting of surfaces and absorption.

1.3.1 Methodology

The rainwater harvesting system employed in this study comprises two basic components: rain collecting surface and storage tank. For collection and storage of rainwater at Coimbatore, the sheet is employed as catchment surface, and rainwater is utilized through processes of collection, purification and storage.

In order to drive DC geared motors to the desired position, the theoretical value of azimuth and altitude angles is transformed into digital commands. Based on the digital commands, the system automatically eliminates the azimuth and altitude angle of the solar PV panel based on feedback signal of the LDR sensor module.

The rain sensor consists of nickel-coated lines, and it works on the principle of resistance. The resistance between the contact lines will be high in the absence of rain, so there is conduction between the wires. In the presence of rain, the rainwater droplets would act as a conduction path between the wires, which will reduce resistance between the contacts. Thus, the analog signal is converted to digital signal and is supplied to the controller for appropriate operation.

The DC series wound motor is coupled to the driving rod, which rotates in forward and reverse direction. A metallic disk is attached to the driving rod (as shown in Fig. 1.3). The supporting rod is a beam which connects the metallic disk and beams which hold the sheet (collecting surface). The supporting rod enables the opening and closing operations, depending upon the movement of metallic disk. When the controller receives the rain detection signal, it drives the DC series wound motor. When the motor (along with driving rod) rotates in forward direction, the metallic disk moves downward. As it moves downward, the sheet opens with the help of supporting rods, as shown in Fig. 1.2.

In the absence of rain, the motor rotates in the reverse direction, and the operation gets reversed. The metallic disk moves upward, and the sheet gets back to its initial

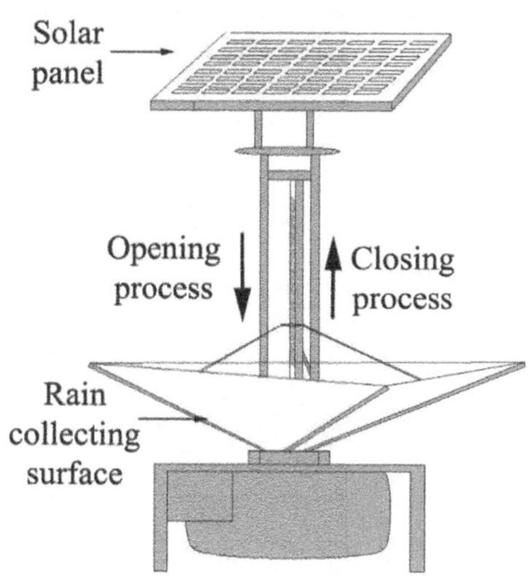

Fig. 1.2 Front view of the intelligent stand-alone solar PV-enabled rainwater collector and power generator

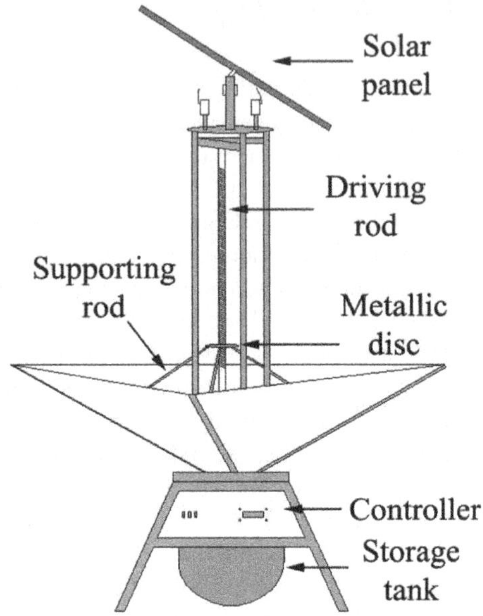

Fig. 1.3 Side view of the intelligent stand-alone solar PV-enabled rainwater collector and power generator

position. The closing of the collecting surface is done in order to avoid the physical contamination such as bird droppings, leaves or any other dirt material.

For generating large amount of energy from solar panel, a high-power panel is required. As solar panel power rating is proportional to its size, if we apply large panel in the proposed system, it will cover up the area of the collecting surface. In order to overcome this problem, the solar panel can be connected separately, as shown in Fig. 1.5. By adding the higher power solar panels, the efficiency of the proposed system can be optimized effectively. The working principle of the above system is similar to the proposed system. The size of the solar panel is independent of the structure. The size and number of the solar panels can be varied depending

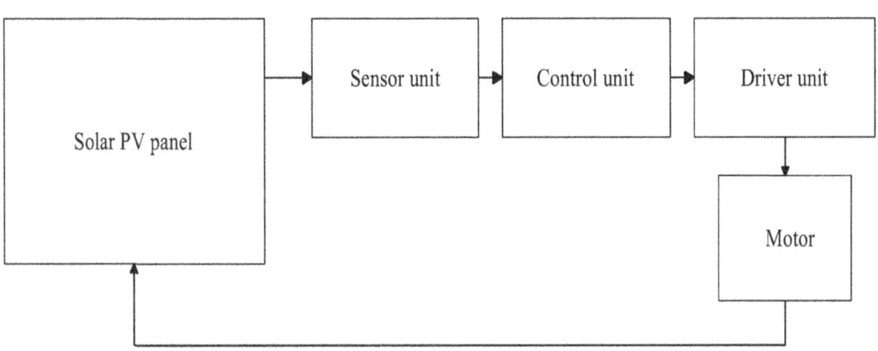

Fig. 1.4 Block diagram of solar tracking system

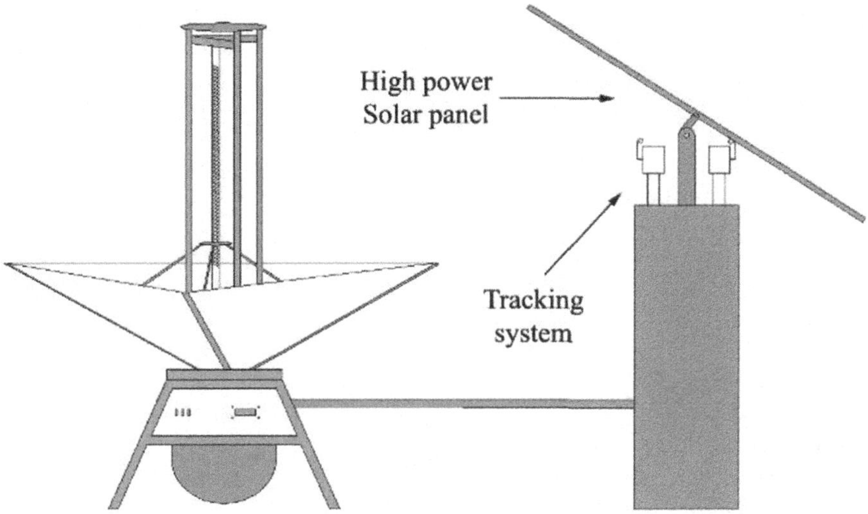

Fig. 1.5 Schematic diagram of proposed system with separate solar PV panel

upon the necessity of the system. This system requires an external additional space for installation of the solar panels. The distance between the rain collection surface and the solar panel can be varied depending upon the location of installation.

1.3.2 Analysis

The 100 W solar panels were examined under two conditions. At first, the solar panel was fixed at a particular angle and then the solar panel was placed on the solar tracking system. At fixed position, panel was placed at the local latitude angle (11°). The test was conducted on May 2019 in Coimbatore, India (longitude 76° and latitude 11°). Temperature in the location varied from 25 to 37 °C. The measurement was taken every 30 min.

Table 1.1 illustrates the data of test conducted, while the solar panel was kept at non-tracking/fixed position.

Under fixed position, the power generated in the solar module varies accordingly with the incident angle of solar radiation.

Table 1.2 illustrates the data of test conducted, while the solar panel was placed on a tracking system. The angle of solar panel varies according to the intensity of sunlight and incident angle of solar radiation.

From Tables 1.1 and 1.2, it is evident that the voltage value measured by the solar panel under tracking system is higher as compared to that of the fixed configuration. Based on the current value, the observed value of solar panel under tracking system is not ably more than the current value of fixed system. As a result of voltage and

Table 1.1 Test data for fixed system

Time	Current (I)	Voltage (V)	Power (W)
7:30	1.3	5	6.5
8:30	1.6	6	9.6
9:30	2.3	13	29.9
10:30	2.4	15	36
11:30	2.4	15.5	37.2
12:30	2.4	15.5	37.2
13:30	2.3	15	34.5
14:30	2.2	13	28.6
15:30	1.4	4	5.6

Table 1.2 Test data for tracking system

Time	Current (I)	Voltage (V)	Power (W)
7:30	1.9	7	13.3
8:30	2.4	15	36
9:30	2.4	16	38.4
10:30	2.5	17	42.5
11:30	2.6	17	44.2
12:30	2.5	16.5	41.2
13:30	2.5	16	40
14:30	2.4	15	36
15:30	2	10	20

current values observed, the power value recorded by the automatic tracking system is also higher than that of fixed system. Therefore, the conversion power of the solar panel can be significantly increased by the automatic tracking system.

Figure 1.6 illustrates the power generated (in kWh) during the year 2019. A

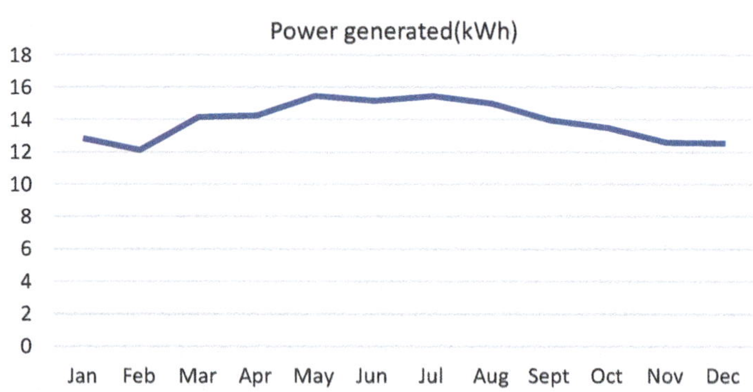

Fig. 1.6 Power generated during 2019

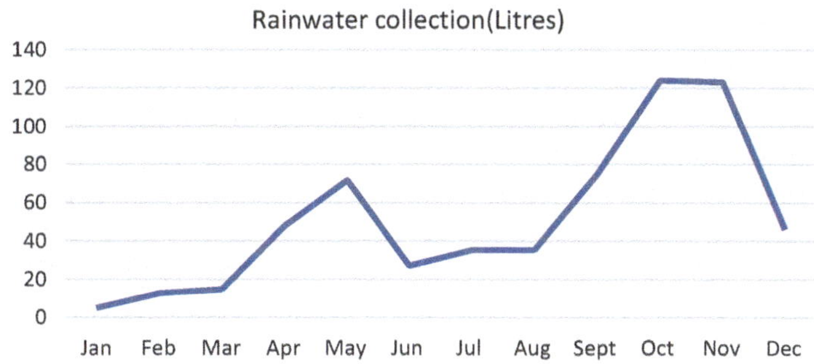

Fig. 1.7 Rainwater collected during 2019

cumulative power generated by the solar modules employed in the system throughout the year is approximately 167 kWh. The maximum power was generated during May, June and July. The power generated during this period is approximately 46 kWh. The lowest power generation was witnessed during the month of February (12.18 kWh).

In 2019, Coimbatore received a total of 616.7 mm rainfall. Coimbatore received least rainfall during the month of January and February. During the month of March, April and May (summer shower season), the city received 134.3 mm rainfall. During June, July, August and September (southwest monsoon season), rainfall of 172.7 was observed. During October, November and December (northeast monsoon season), 291.3 mm rainfall was detected. Coimbatore received a total of 54 rainy days.

Figure 1.7 illustrates the amount of rainwater collected during 2019. Maximum amount of rainwater was collected during the northeast monsoon season as this season received the highest rainfall in the year. Approximately, 295 L of rainwater had been collected during this season. After northeast monsoon, a fair amount of rainfall received was during summer shower season. Approximately, 136 L of rainwater had been collected during this season. The accumulation of the rainwater in the year 2019 in Coimbatore is estimated about 625 L from this system.

Maximum power consumption of this device is during the operation of the main motor, and the main motor is activated in response to the rain sensors for opening and closing operations of the device. Therefore, battery backup of the system depends upon the continuity of the rainfall.

With respect to the battery installed in this system, if it rains twice a day, the device has the capability to last for approximately 30 days without recharging. This may vary according to the climatic condition of a certain place. In extreme conditions where the availability of the sunlight is very low or negligible for many days, the battery can be recharged externally or can be replaced with another battery.

1.4 Results and Discussion

An intelligent stand-alone solar PV-enabled rainwater collector was constructed by utilizing resources and materials locally available and was found to be highly effective. Collection of rainwater can be considered as green building practices as it is eco-friendly and is useful in avoiding flooding. The total amount of rainwater collected in a year is approximately 625 L as shown in Fig. 1.7. It is evident that the amount of rainwater harvested could be utilized effectively during dry periods of the year.

The total power generated by the system is approximately 167 kWh as shown in Fig. 1.6, which is capable of driving the system as well as providing supply to external loads.

Since the scrutinized rain harvesting technology does not have any ramification on the environment, it appears to be a sustainable and efficient method. Therefore, advocacy for the espouse of rainwater will result in reduction of constraints related to shortage of water in parch cities.

References

1. X. Wang, An applied research in the rural landscape of rainwater collection system based on the concept of LID, in *International Conference on Intelligent Transportation, Big Data & Smart City* (2015), pp. 854–857
2. B. Kumar Biswas, B. Hira Mandal, *Construction and Evaluation of Rainwater Harvesting System for Domestic Use in a Remote and Rural Area of Khulna, Bangladesh.* International Scholarly Research Notices (2014)
3. P. Bitterman, E. Tate, K.J. Van Meter, N.B. Basu, Water security and rainwater harvesting: A conceptual framework and candidate indicators. Appl. Geogr. **76**, 75–84 (2016)
4. M.F. Colom-Reyes, A. Soriano-Gómez, J.M. Hernández-Martínez, M. Trejo-Perea, O. Chávez-Alegría, G.J. Ríos-Moreno, *Rainwater Harvesting for Household Use*, XIII International Engineering Congress (CONIIN), Santiago de Queretaro, Mexico (2017)
5. S. Raghavan, *Rainwater Harvesting—The Success Story of Chennai*, vol. 13, IEEE India Info. (2018), pp. 30–34
6. G. Zhang, K. Yang, F. Wu, S. Yang, N. Li, Q. Zhu, Introduction of rain harvesting treatment in Northwest Villages of China, in *International Symposium on Geometrics for Integrated Water Resource Management* (2012)
7. A. Morey, B. Dhurve, V. Haste, B. Wasnik, Rain water harvesting system. Int. Res. J. Eng. Technol. **3**, 2158–2162 (2016)
8. H. Traboulsi, M. Traboulsi, Rooftop level rainwater harvesting system. Appl. Water Sci. (2017)
9. D. Shah, P. Patel, P. Kudo, A. Steinberg, W. Herold, Feasibility of rainwater harvesting on Scott Laboratory, in *IEEE 2013 Global Humanitarian Technology Conference* (2013)
10. C. Sharma, Dr P. Thenkabail, J.R. Sharma, Earth observing data and methods for advancing water harvesting technologies in the semi-arid rain-fed environments of India, in 2011 IEEE Global Humanitarian Technology Conference (2011)
11. Y.-R. Chiu, Y.-L. Tsai, Y.-C. Chiang, Designing rainwater harvesting systems cost-effectively in an urban water-energy saving scheme by using a GIS—simulation based design system. MDPI **7** (2015)

12. H. Qin, Application of rainwater regulation (collection) storage system technology for landscape green lands & squares in Tianjin City, in *International Conference on Mechatronics and Automation* (2012)
13. J. Zankowski, Y. Sun, C. Poon, E. Pas sauer, A. Nassar, K. Mehta, Gutter design and selection for roof rainwater catchment systems, in *IEEE Global Humanitarian Technology Conference* (2013)

Chapter 2
Speed Control of SEPIC Converter-Based Induction Motor Drive System

R. Meena Devi, V. Geetha, and V. Meenakshi

Abstract This paper deals with speed control of single-ended primary inductance converter fed induction motor drives for water pumping application. A SEPIC converter with passive component is designed for PV system that provides regulated output voltage. In this, regulated voltage is used to drive induction motors in a wide range of speed. In closed loop, fuzzy logic control system is used to generate the switching pulse for SEPIC converter. Pulsating input current, high voltage stress of conventional method will make the unreliable for a wide range of speed. So, the SEPIC converter is used to for smooth speed control process. Simulation is down in MATLAB, Simulink. And then, implemented PV for induction motor drives is compared with theoretical values. SEPIC converter is well suitable for smooth speed control of induction motor. The SEPIC also has a simple controller that provides low noise operation.

Keywords Induction motor · SEPIC converter · Fuzzy controller · Speed control

2.1 Introduction

Solar power is more and more attractive due to non-polluting in nature and best alternative for the conventional energy sources, which is declining. Photovoltaic system has a large number of arrays are connected in series and parallel combination that the cell temperature produces arrays of voltage. Each PV array has maximum operating point. And the location of the PV and the angle of the PV array are important to find the operating point.

A maximum power point tracker is needed to operate the PV array at its maximum power point. In this research, a technique for extracting the maximum output power from an efficient PV system is proposed by a single-ended primary inductor converter (SEPIC), simplified multilevel inverter and a voltage source inverter (VSI) for power quality improvement.

R. Meena Devi (✉) · V. Geetha · V. Meenakshi
Sathyabama Institute of Science and Technology, Chennai, India

© The Author(s), under exclusive license to Springer Nature Singapore Pte Ltd. 2021

V. Kamaraj et al. (eds.), *Emerging Solutions for e-Mobility and Smart Grids*,
Springer Proceedings in Energy, https://doi.org/10.1007/978-981-16-0719-6_2

India is an agricultural country. It is dependent on electricity for water pumping application. Nowadays, research is focussing on PV for rural area. So, the PV SEPIC converter fed induction motor. Solar power is produced from sunlight and converted into a small amount voltage. This voltage can be stepped up into higher value of voltage using single-ended primary inductance converter (SEPIC). VSI is used to convert the variable DC voltage into AC voltage. This voltage is used to drive the induction motor. Induction motor is well suitable for fan water pumping and lightning.

2.2 SEPIC Converter with Induction Motor Drive for Open Loop System

The circuit diagram for SEPIC converter with induction motor is as shown in Fig. 2.1. When light energy falls on PV panel, it converts light energy into electrical energy, i.e., voltage, the SEPIC converter used to boost up the voltage, the inverter converts the DC voltage into three-phase AC voltage, the motor rotates, and therefore, we can see the improvement of speed in the induction motor using SEPIC converter (Fig. 2.2).

The output speed, voltage and current waveform for SEPIC converter with induction motor is as shown in Figs. 2.3 and 2.4. THD is shown in Fig. 2.5. It shows that SEPIC converter has induced low value of total harmonic detection (THD) in input current.

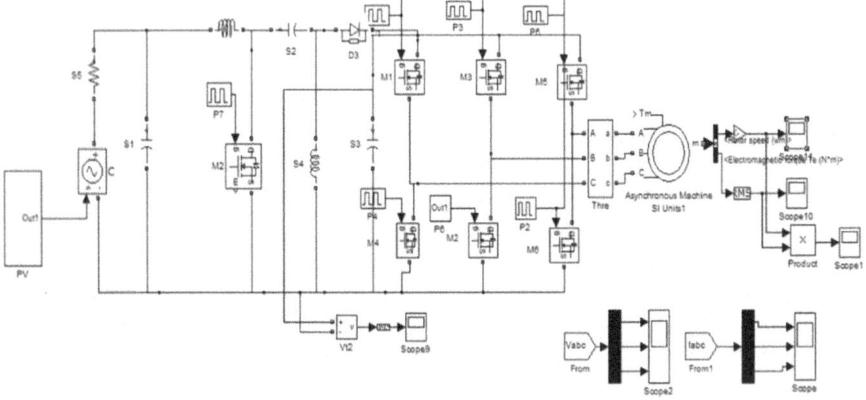

Fig. 2.1 Open loop control of SEPIC converter fed inductance motor drive

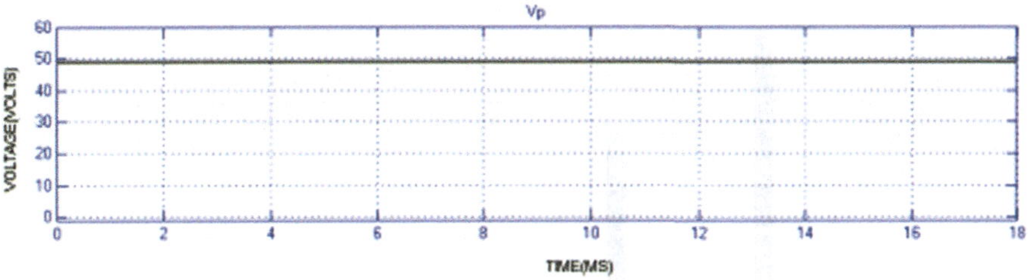

Fig. 2.2 Input voltage for SEPIC converter

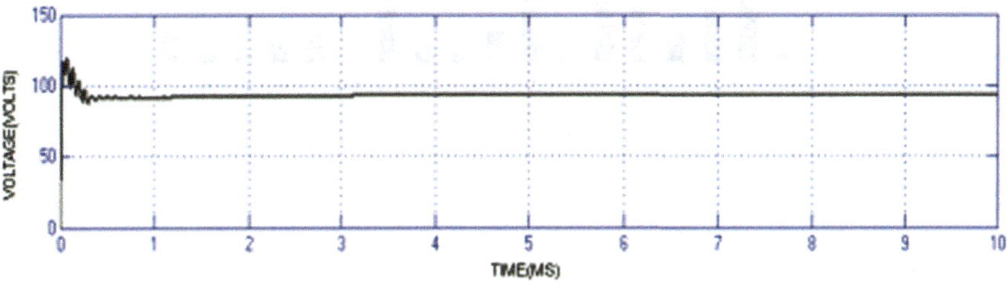

Fig. 2.3 Voltage across SEPIC converter

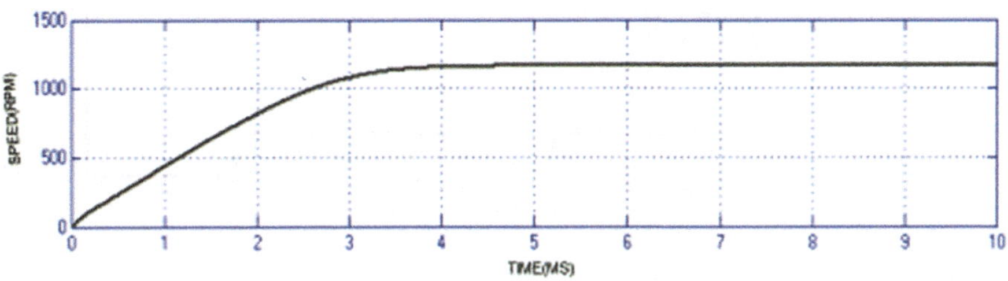

Fig. 2.4 SEPIC converter Motor speed

2.3 SEPIC Converter with Induction Motor for Closed Loop System

2.3.1 Proposed Block Diagram

Photovoltaic impact is a fundamental physical procedure through which sun-oriented vitality is changed over into electrical energy specifically. The material science of a PV cell, or sunlight-based cell, is like the traditional PN intersection diode. During the evening, a PV cell can essentially be considered as a diode (Fig. 2.6).

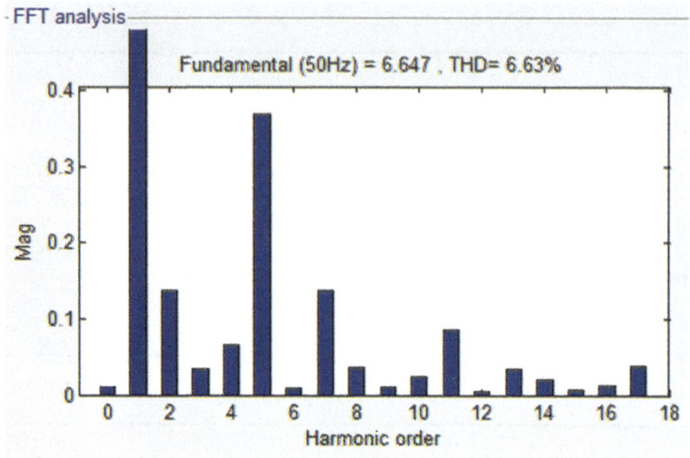

Fig. 2.5 SEPIC converter THD

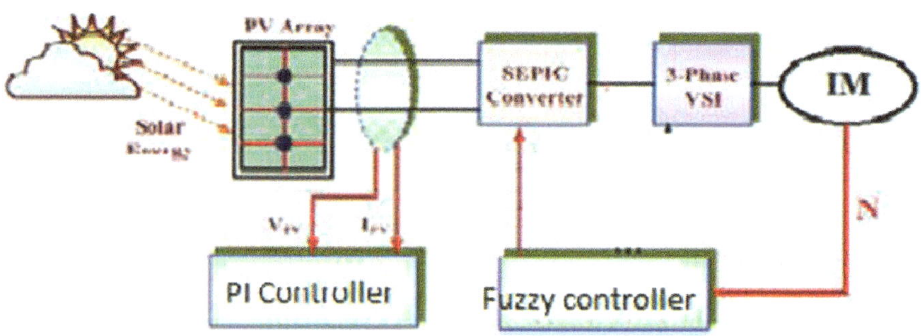

Fig. 2.6 Block diagram for proposed SEPIC converter

2.3.2 *Proposed System Simulation Diagram*

The proposed system is shown in Fig. 2.7. It consists of two components as inputs, one is PV panel, and another is battery. It also consists of SEPIC converter, three-phase inverter and an induction motor. In this, the closed loop is used, i.e., the feedback is taken from the output, and it is given as input. When the light falls on PV panel, it converts light energy into electrical energy, i.e., the voltage, the SEPIC converter is used to boost up the voltage, the voltage is given to the inverter, the inverter converts DC voltage into AC voltage and is given to motor, and the motor rotates, the reference speed is taken from the induction motor and is given to the fuzzy logic controller, and the output will be accurate because of fuzzy logic controllers, the steady state errors are used (Figs. 2.8 and 2.9).

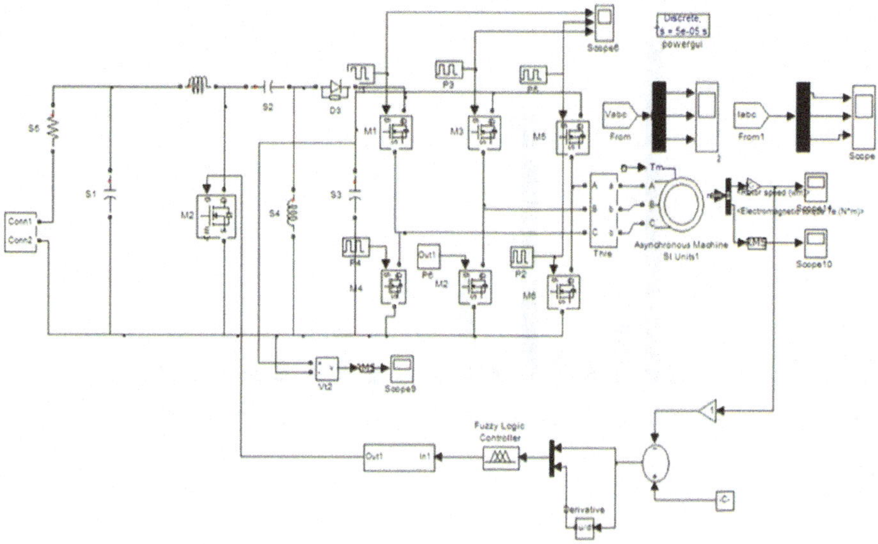

Fig. 2.7 Proposed SEPIC converter with induction motor

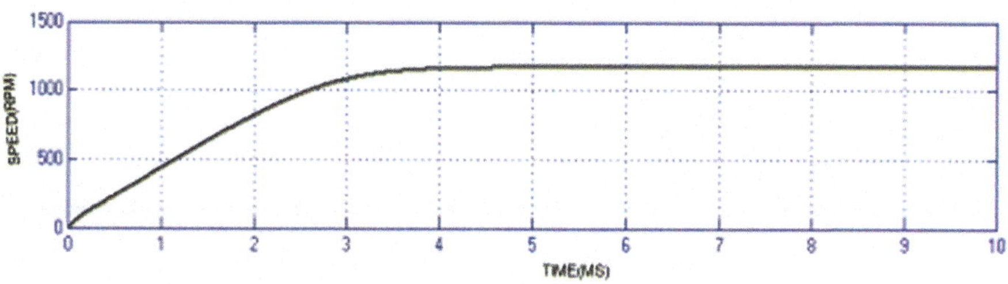

Fig. 2.8 Proposed SEPIC converter with induction motor speed

2.3.3 *SEPIC Converter with Induction Motor with Fuzzy Controller*

A fuzzy logic control system is used to control pulse for SEPIC converter system based on fuzzy logic a mathematical system that analyzes. The FLC system has analog input values in terms of logical variables which operates on discrete value either 0 or 1. The proposed system surface area and rule of SEPIC converter are shown in Fig. 2.10a, b.

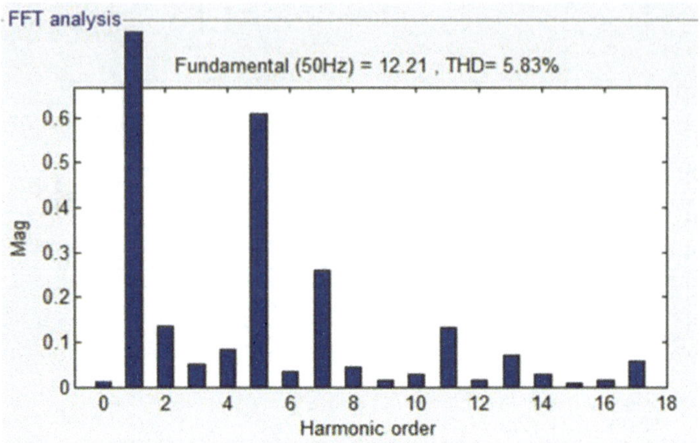

Fig. 2.9 Proposed SEPIC converter THD

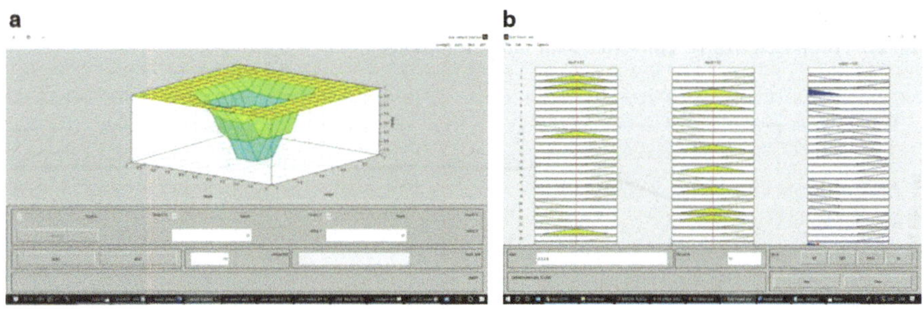

Fig. 2.10 **a** Surface of fuzzy controller file, **b** rules for SEPIC converter file

2.4 Hardware Implementation SEPIC Converter with Induction Motor for Closed Loop System

When light energy falls on PV panel, it converts light energy into electrical energy and it is shown in Fig. 2.11. Implementation of proposed system with induction motor is shown in Fig. 2.12. The SEPIC converter boosts up the voltage, and it is passed through the inverter; the inverter converts DC voltage into three-phase AC voltage, it is given to motor, the motor rotates, and hence, we will observe the improvement of speed using SEPIC converter (Table 2.1).

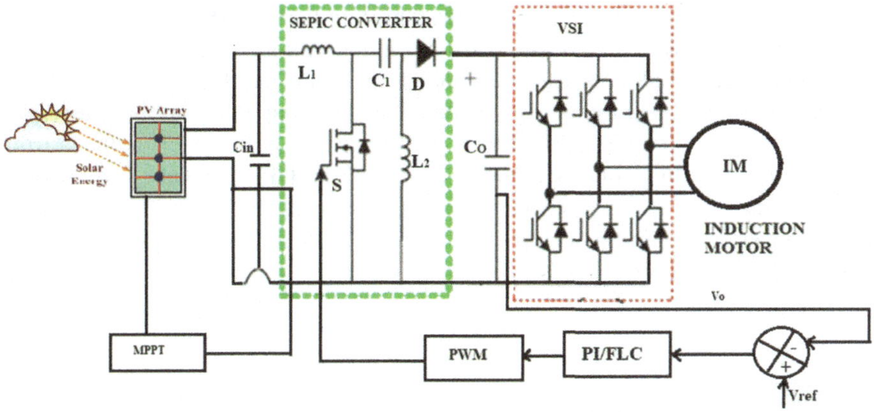

Fig. 2.11 Proposed SEPIC converter hardware block diagram

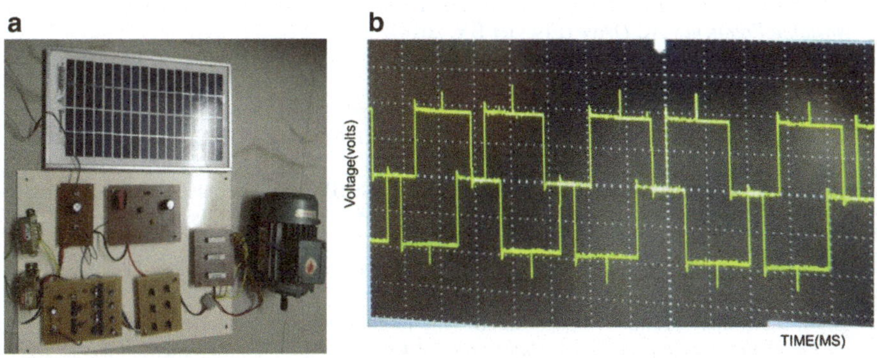

Fig. 2.12 **a** Hardware setup, **b** inverter output voltage of induction motor drive

Table 2.1 Table open loop and closed system with Induction motor drive

PV with three-phase inverter	Vo (V)	*N* (RPM)	THD (%)	Po (W)
Open loop SEPIC	63	950	6.63	600
SEPIC	92	1160	5.83	1000
Experimental SEPIC	90	1100	6.8	1000

2.5 Conclusion

SEPIC converter with an induction motor drive and boost converter with induction motor drive are simulated, and outcomes are presented. By using the SEPIC converter, the output voltage is increased from 63 to 92 V; motor speed is increased from 950 to 1160 RPM; THD is reduced from 6.63 to 5.83%, and output power is increased from 600 to 1000 W. Hence, the outcomes represent that the SEPIC converter with induction motor drive is superior to boost converter with induction motor drive.

PI controlled SEPIC converter with induction motor and FL controlled SEPIC converter with induction motor are simulated, and the outcomes are presented. By using FLC, the rise time is reduced from 1.07 to 1.01 s; peak time is reduced from 1.13 to 1.05 s; settling time is reduced from 1.32 to 1.11 s; steady state error is reduced from 6.7 to 1.8 V. Hence, the outcomes represent that the FL controlled SEPIC converter with induction motor is superior to PI controlled SEPIC converter with an induction motor.

References

1. S. Kaliappan, R. Rajeswari, An investigation of power converters fed BLDC motor for adjustable speed. Circ. Syst. (2016)
2. S.K. Dwivedi, *Isolated Active Waveshaping Techniques for Power Quality Improvements in Variable Frequency AC Drive* (Elsevier BV, 2018)
3. S. Malathi, Improved power quality buck boost converter for SMPS. Int. J. Electr. Comput. Eng. **9**(2), 789–801 (2019)
4. X. Chen, G. Liu, Sensorless optimal commutation steady speed control method for a non-ideal back-EMF BLDC motor drive system including Buck Converter. IEEE Trans. Ind. Electron. (2019)
5. S. Gao, Y. Wang, Y. Guan, D. Xu, A high step up SEPIC-based converter based on partly interleaved transformer. *A High Step Up SEPIC-Based Conics*, vol. 67, Issue 2, pp. 1455–1465 (2020)
6. R. Meena Devi, Fuzzy logic based sensorless speed control of SEPIC Fed BLDC drive. Int. J. Appl. Eng. Res. **10**(2), 2715
7. R. Meenadevi, V. Geetha, PV fed bridgeless DC motor using zeta converter for water pumping, in *Proceedings of the 2019 2nd International Conference on Power and Embedded Drive Control, ICPEDC 2019* (2019)
8. R. Meena Devi, L. Premalatha, A new PFC converter using bridgeless single-ended primary induction converter(SEPIC), in *2015 International Conference on Control Instrumentation Communication and Computational Technologies, ICCICCT 2015*
9. R. Meena Devi, L. Premalatha, Efficient figure converter fed PMBLDC motor using artificial neural network. Int. J. Electr. Comput. Eng. **9**(4), 3025–3031 (2019)
10. A. Kavitha, G. Uma, Control of Choos in SEPIC DC-DC converter. Int. J. Control Autom. Syst. **8**(6), 1320–1329 (2010)
11. C.G. Bianchin, R. Gules, High-power-factor rectifier using the modified SEPIC converter operating in discontinuous conduction modes. IEEE Trans. Power Electron. **30**(8), 4349 (2015)
12. K.J. Rathi, M.S. Ali, NNC for power electronics converter circuits. Int. J. Electr. Electron. Res. **4**(1), 78–84 (2016)
13. L. Yang et al., Investigation of a stator-ironless brushless DC motor with non-ideal back-EMF. Digital Object Identifier https://doi.org/10.1109/access.2019.2901632
14. Y. Li et al., A sensorless commutation error correction method for high-speed BLDC motors based on phase current integration. IEEE Trans. Industr. Inform. **16**(1) (2020)
15. S. Sashidhar et al., A single-stage sensorless control of a PV-based bore-well submersible BLDC motor. IEEE J. Emerg. Sel. Topics Power Electron. **7**(2) (2019)

16. R. Meena Devi, M. Kavitha, V. Geetha, M. Preethi Pauline Mary, Fuzzy logic controller-based bridgeless Cuk converter for power factor correction, in *Advances in Intelligent Systems and Computing*, vol. 846 (AISC, 2018), pp. 371–379
17. R. Meena Devi, L. Premalatha, Analysis of bridgeless converter model for power factor correction. Comput. Electr. Eng. **87**, 106785 (2020)

Chapter 3
The Hybrid Electric Vehicle (HEV)—An Overview

E. Fantin Irudaya Raj and M. Appadurai

Abstract The environmental impact created by automobiles is becoming a social issue day by day. To overcome these issues and save the future from worse alternative technologies like hybrid electric vehicle (HEV) and electric vehicle (EV) are required and introduced in the automobile industry. EV helps us to attain a 100-percentage clean in service. But it has infrastructure limitations and short-driving-range problems. To overcome this, we are in need of a hybrid system. HEV is a beautiful substitute for the traditional internal combustion (IC) engine-based vehicle system and reduces the problems caused by emission. It provides a suitable solution in the aspect of infrastructural limitation and reduction in operating costs. As the name suggests, HEV is a combination of IC engine vehicle and electric vehicle. Whereas IC engine vehicle runs on gasoline product and electric vehicle runs by an electric motor. Also, In HEV, the electric motor is connected with a battery pack which is rechargeable for electric motorized driving. At the same time, HEV uses both engines to increase the power and torque or depending upon the driving type, rely on anyone. This paper discusses the recent advancement of HEV, various factors involving, the challenges and the state-of-the-art solutions by automobile industries and academicians in manufacturing.

Keywords Battery vehicles · Hybrid vehicles · Plug-in hybrid vehicles

3.1 Introduction

The unceasing progress of our world economy, the rise of population and enhancement of people's lifestyle leads to the increment in utilizing and consumption of energy. This will also lead to environmental pollution and global warming [1]. Most

E. Fantin Irudaya Raj (✉)
Department of Electrical and Electronics Engineering, Dr. Sivanthi Aditanar College of Engineering, Tiruchendur, India

M. Appadurai
Department of Mechanical Engineering, Dr. Sivanthi Aditanar College of Engineering, Tiruchendur, India

© The Author(s), under exclusive license to Springer Nature Singapore Pte Ltd. 2021
V. Kamaraj et al. (eds.), *Emerging Solutions for e-Mobility and Smart Grids*,
Springer Proceedings in Energy, https://doi.org/10.1007/978-981-16-0719-6_3

of the countries have been discussing the adverse effect of global warming since the early part of the twenty-first century. Many pertinent studies have reported the aggressive impact of climate change caused by human activities. All because of the global level increase in civilization and industrialization leads to a large quantity of fossil fuel burnings through transportation, resulting in air pollution and global warming [2]. The exhaust emissions from vehicles are the major contributor to the impact of greenhouse gases. Emissions mainly include CO, CO_2 and particulates like PM2.5 & PM10. These are also the main reason behind the lung cancer and other serious respiratory diseases.

Development in the transportation sector captivates about 50% of oil resources worldwide and the main reason behind the depletion of renewable energy sources. Therefore, it is a great need to substitute non-renewable energy resources and use adequate energy-saving technologies. There is a great need to develop alternative energy sources to improve heat energy conversion efficiency to solve environmental pollution [3, 4].

Electric vehicles (EVs) have been systematically investigated and considered a potential solution to transport-related environmental problems [5]. Based on the propulsion devices and the power supplement, EV is classified into three types: (1) PEV—pure electric vehicle or battery electric vehicle (BEV), (2) HEV—hybrid electric vehicle (HEV) and (3) FCEV—fuel cell electric vehicle (FCEV). The PEV is nourished from the power storage unit solely by electrical energy, and it is propelled exclusively by an electric motor. HEV's working mechanism includes the electrical motor and IC engine, whereas electric energy from the battery system and gasoline or diesel is used as the power source. FCEV is using an electrical motor for its propulsion, and it could be powered by a fuel cell [6]. Brief descriptions of various EVs, storage facilities, charging EVs via PVs, other socio-technical dispute for EV adoption and the global potential of EVs have been highlighted in this review paper. A rise in EV adoption relies on government policies that offer rewards and benefits that are lucrative [7].

3.2 Hybrid Electric Vehicle

Vehicle emission is a central problem for climate change policy. The advancements in vehicle efficiency can reduce fuel consumption and vehicle harmful discharges. The usage of hybrid vehicles and plug-in hybrid vehicles are the available techniques used to reduce greenhouse gases. The hybrid vehicles enhance the fuel economy, and the plug-in hybrid vehicles utilize the electricity for partial propulsion.

A hybrid vehicle utilizes at least two different power sources, such as fossil fuels, electricity, etc. The main principle in a hybrid vehicle is that different power sources utilize at different speeds. The motor produces higher torque efficiently, and the IC engine can produce higher speed (compare to an electric motor). Utilizing both efficient outputs becomes energy efficient and less fuel consumption.

The hybrid vehicle efficiency is improved by the advanced control systems, and its performance in different load is enhanced by technologically advanced control systems. A proper advanced controller is necessary for efficient management of the energy. The DC-DC boost power converter is used to maintain a constant voltage. The energy released during regenerative braking is used to recharge the batteries. Table 3.1 denotes the different types of vehicles—a comparison and Table 3.2 represents the advantages, disadvantages and application range of different type of vehicles.

Table 3.1 Different type of vehicles—a comparison

Type of vehicle	Drive unit	Energy sources	Energy supplements
PEV	Electrical motor	Battery source, ultracapacitor	Electrical energy from power system or battery
HEV	Electrical motor and IC Engine	Battery, ultracapacitor, IC engine	Gasoline station and electrical energy from power system or battery
FCEV	Electrical motor	Hydrogen and oxidizing agent	Fuel cells
ICEV	Internal combustion engine	Gasoline or diesel	Gasoline station

Table 3.2 Advantages, disadvantages and application range of different type of vehicles

Type of vehicle	Advantage	Disadvantage	Application range
PEV	Pollution-free, low cost and need a special infrastructure for charging	Short driving distance and huge battery pack is needed	Suitable for low-speed and short-range community
HEV	Hybrid technology and Suitable to adopt in the current infrastructure itself	High cost	Meet out the daily needs. Suitable for high-speed and long-range community
FCEV	Advanced technology and need a special infrastructure	Costlier than PEV, HEV and ICEV	Long-range and high-speed community. Provides more mileage compared with other types of vehicles
ICEV	Existing in more numbers, no need for special infrastructure	Burning of fuels leads to environmental pollution	Mostly used in all range vehicles and suitable for high speed to low speed

3.3 Classification of Hybrid Electric Vehicle

HEV is classified based upon the construction, hybridization and refueling methods. Based on the structure (Fig. 3.1), the HEV is further classified into three different categories. They are (i) series hybrid configuration, (ii) parallel hybrid configuration and (iii) series–parallel hybrid configuration.

The series hybrid configuration is shown in Fig. 3.2. This type of vehicle is driven only by the electric motor. The motor itself acts as a generator at the time of regenerative braking. This construction streamlines the power train design because clutch and reduction gear both are not necessary. By controlling the electric motor alone, we can control the speed and torque. In this configuration, the role of the IC engine is charging the battery through the generator and supplying energy to the electric motor, provides chances to operate in maximum efficiency. This is one of the strategies that help to increase overall efficiency. Therefore all because of these reasons, the series hybrid vehicle is also known as an IC engine-assisted electric vehicle. The main drawback of this configuration is it needs one IC engine, generator and motor.

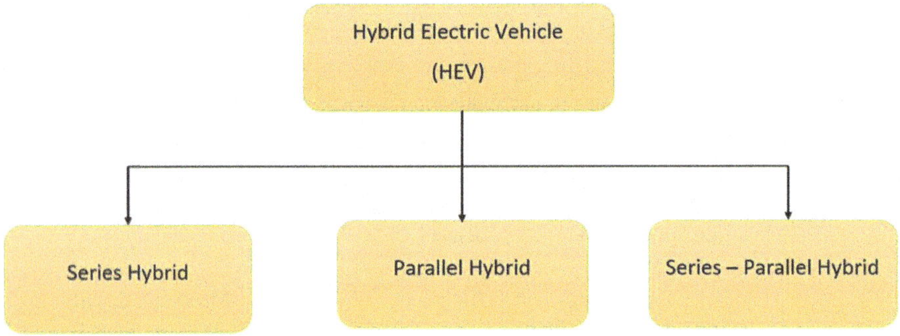

Fig. 3.1 Classification based upon the structure of HEV

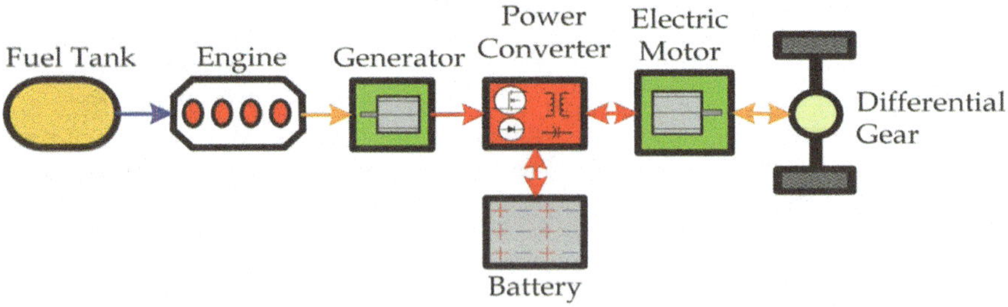

Fig. 3.2 Series hybrid configuration [8]

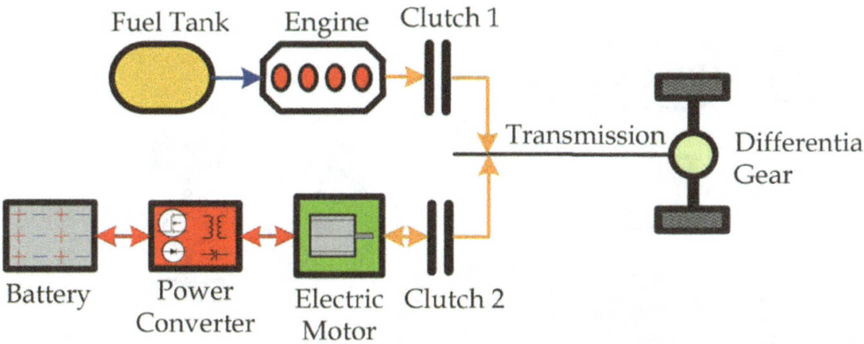

Fig. 3.3 Parallel hybrid configuration [8]

Hybrid electric vehicles' propulsion with an electric drive mechanism and IC engine is continuously boomed for their energy-efficient, low emission and recoverable regenerative braking system. In a series connection, the engine is disengaged from the mechanical propulsion system to operate in optimum efficiency for better performance. Thus, the losses of electrical and mechanical energy are minimized for system performance improvement.

Figure 3.3 shows the parallel hybrid configuration. In this type of HEV, the propulsion of the vehicle will be the result of torque produced through the simultaneous operation of ICE and the electric motor. This methodology provides the opportunity to use the IC engine and electric motor independently by using two clutches. This topology is superior through providing greater dynamic performance compare with series hybrid configuration. Parallel HEV is also a known electric motor-assisted IC engine vehicle. The parallel hybrid configuration is more suited for the costliest car segment and full hybrid segment. Thus, parallel connection has both electric motor and engines engaged in the mechanical propulsion path. Owing to its uncontrollable speed, the engine efficiency and the energy utilization are lowered.

The series–parallel hybrid is a mixture of both series and parallel hybrid configurations, which is represented in Fig. 3.4. Compared with the other two configurations, this one will be expensive. It needs planetary gear and another one electric generator. This model is preferred when automobile engineers target high cruising speeds and good dynamic performance.

Based upon the hybridization, the HEV is further classified into another three categories. They are (1) micro-hybrid, (2) mini hybrid and (3) full hybrid.

This mode of classification is based upon the hybridization degree of HEV denoted in Fig. 3.5. Similarly, it specifies the importance of the electric motor which is used in the vehicle propulsion. Where micro-hybrid uses an electrical motor of about 2.5 kW at 12 V, in this arrangement, the electric motor is operating at the start and stopping condition, mainly used in the city or high traffic region driving. In this way, the electric motor is acting as a helping hand to the IC engine. In this vehicle configuration, energy saving is about 5–10%. It provides the worst economy, with insignificant impression on fossil fuel requirements. Mild hybrid uses electric motor

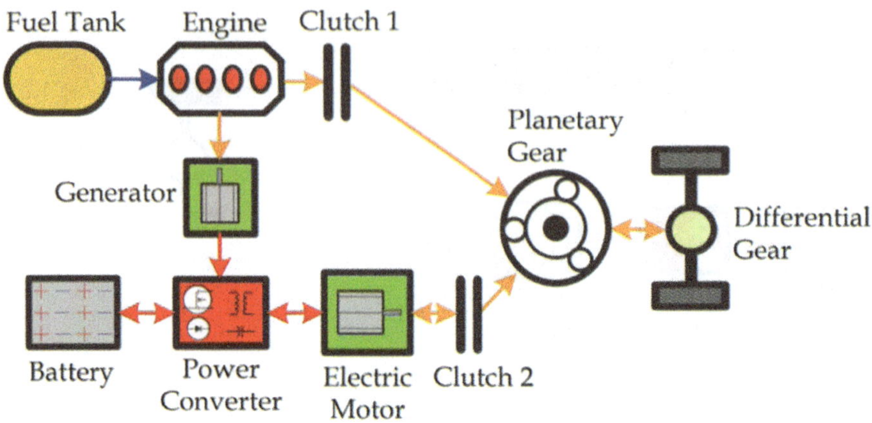

Fig. 3.4 Series–parallel hybrid configuration [8]

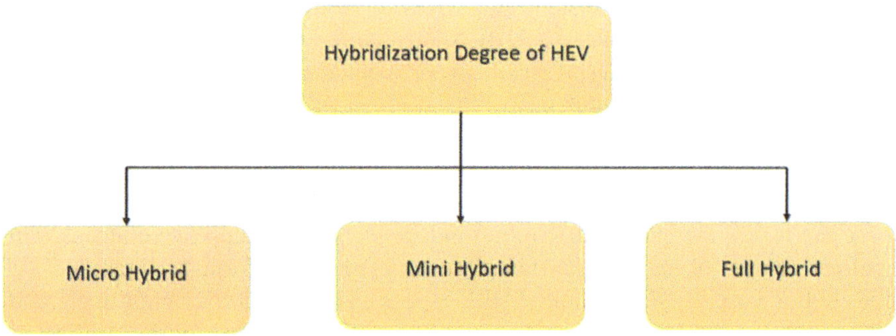

Fig. 3.5 Classification of HEV based upon hybridization

about 10–20 kW at 100–200 V. In this configuration, The IC engine and the electric motor drive the vehicle based upon the condition of the vehicle (start or stop or increasing or decreasing). In this vehicle configuration, energy saving is about 20–30%. It provides somewhat good economy compared with mini hybrids. So many commercial models are in the market based upon this configuration. It is costlier than the typical IC engine vehicle even though its less operational cost attracts more consumers. The full hybrid uses electric motor of about 50 kW at 20–300 V. By using complex control algorithms, the vehicle can manage to operate the IC engine only when needed. It also directs the excess energy to batteries. In this model, the energy saving is about 30–50%.

In a similar way, based upon the refuelling methods, the HEV can be further classified into two different categories. They are (i) plug-in HEV (PHEV) and (ii) mild HEV (MHEV).

In PHEV, batteries can be charged by the external source by using a plug directly from the power grid [9]. There are different types of the plug used to charge in various countries based upon the available power supply. In MHEV, there is no plug-in provided to charge the batteries. The IC engine itself is used to recharge the batteries. And so, there is no need for any unique charging stations for this type of vehicle. But compared with MHEV, PHEV is more eco-friendly, and the PHEV provides more fossil fuel independence.

3.4 Electrical Propulsion System

The electrical propulsion system is the heart of HEV. It consists of an electric motor, converter and power electronic controller. The electric motor converts the electric energy into mechanical energy. That energy is used to propel the vehicle. The same motor acts as a generator at the time of regenerative braking and generates electrical power. It is used to charge the battery in the vehicle. The power converter is used to supply a suitable voltage and current to the electrical motor. The power electronic controller which is used to provides the control signal to the converter through controlling the electrical motor to produce required torque and speed. The power electronic controller gets the information through various sensors in the vehicle, and based upon the information received, it produces the suitable control signal to the power converter.

The choice of an electric propulsion system mainly depends upon the driving profile and energy source. The driving profile comprises of maximum speed, acceleration, range, climbing capability and braking. The energy source depends upon the batteries, fuel cells, flywheels and hybrid sources. Compared with industrial applications of motors, the motors used in HEV require frequent starts, stops, acceleration and deceleration. Thus, the electrical motor drive is classified, as shown in Fig. 3.6. Based upon the need of the consumers and technology available with manufacturers, the motor drive got selected. The classification of electrical motor drive is presented in Fig. 3.7.

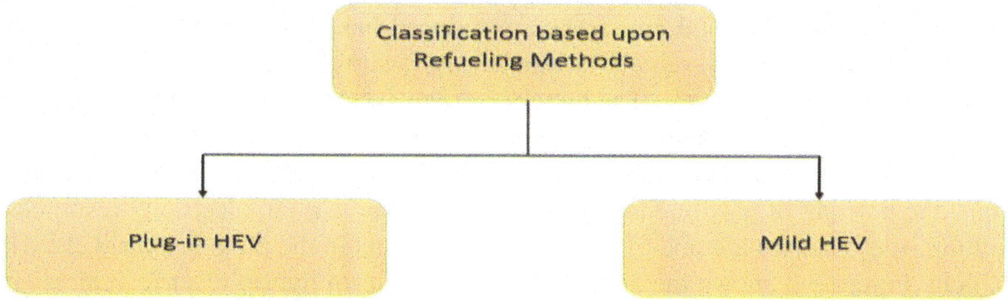

Fig. 3.6 Classification of HEV based upon refueling methods

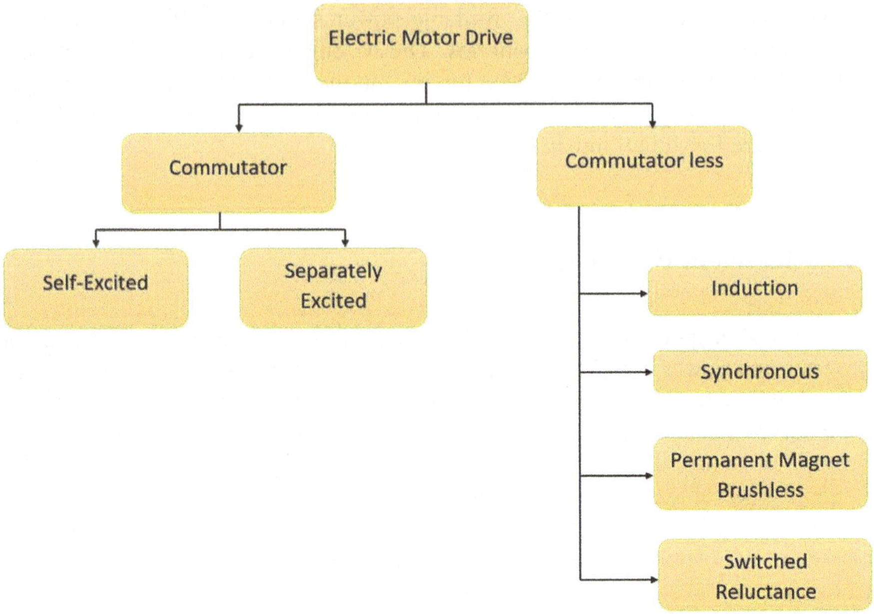

Fig. 3.7 Classification of electrical motor drive

3.5 Advancements of Hybrid Vehicle Propulsion System

The plug-in hybrid electric vehicle requires higher energy density capacity by utilizing lithium-ion battery technology and energizing the vehicle in serial or parallel mode with IC engines. The electrical energy is being used from the grid for charging. These types of vehicles propel the engine using electricity for the specific distance based on the battery storage capacity. The vehicle connected with the electricity supply unit in the non-usage conditions reduces the battery charging on the onboard IC engines. This idea is very attractive for minimizing on-road emissions by using the electric drive. The total emissions are reduced based on electricity generation techniques. Notably, this vehicle is monetarily desirable who have cheaper electrical energy than fossil fuels. The battery storage system consumes the advantages of off-peak time usage benefits from electricity suppliers. Thus, the emissions are minimized when using low-carbon fuel for generating electricity. In on road conditions, the burning of gasoline is reduced by using low-carbon electricity resources. Lithium-ion batteries are one of the main components in the plug-in hybrid electric vehicle. Figure 3.8 shows the state of charge of plug-in hybrid electric vehicles in various conditions.

The series and parallel hybrid engine concepts are having contrary features. For merging two advantages, the parallel concepts integrate with the continuously variable transmission concept [9] are evolved in hybrid engines. This concept allows the hybrid engine to maintain the constant speed, and the motor supplies constant

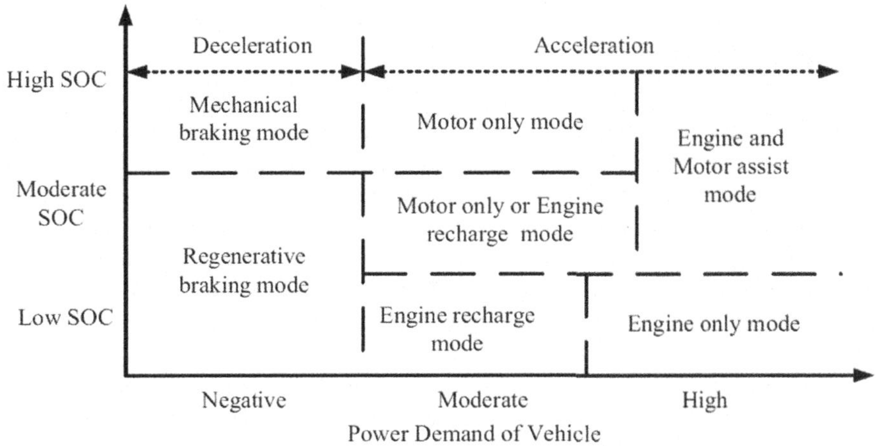

Fig. 3.8 EMS's operation of plug-in hybrid electric vehicle

torque irrespective of the loads and its schematic diagram is shown in Fig. 3.9. And so, the engine efficiency is equal to the series configuration, and the losses on energy conversion are minimum as that of the parallel configuration. A proper cell unit is needed for attaining a reliable energy storage system, and they can utilize its full potential.

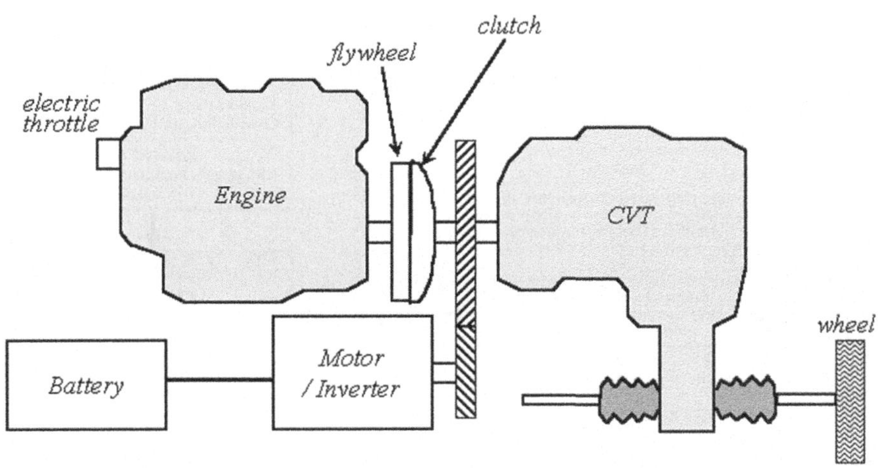

Fig. 3.9 Schematic diagram of hybrid vehicle with continuously variable transmission [9]

3.6 Circumstances Influencing Plug-in Hybrid Fuel Intake and Emissions

The performance and emissions depend upon the strategic energy management by the manufacturer. Primarily how long the electricity is utilized from the battery then after the fuel is utilized in IC engines. While running in electricity, the vehicles have zero emissions. The driver can opt to run the vehicle in pure electric discharge from fully charged batteries. This is helpful to drive the vehicle in emission-free zones when the hybrid vehicle is fully charged before entering the pollution-free zone area. The charging frequency depends on the payload and range of the vehicles to drive. If the parking sites having recharging facilities reduces the exclusive charging time. The energy efficiency is maximized by fully recharging the batteries of hybrid vehicles on each trip, and then the battery gave better performance. The life cycle of hybrid vehicle is shown in Fig. 3.10.

This research provides an in-depth analysis of the different energy management methods for traditional HEV/PHEV and the use of knowledge from vehicle to infrastructure or vehicle to vehicle, including a detailed overview of energy management methods using various methodologies. The energy conservation techniques for HEV/PHEV under an intelligent transport scheme are extensively reviewed with regard to single-vehicle and multi-vehicle scenarios [10].

Hybrid electric vehicles' energy management strategies (EMSs) have been thoroughly researched and evaluated. The offline EMSs target to decrease fuel consumption. While the EMSs cannot be specifically applied in a specific car, other energy

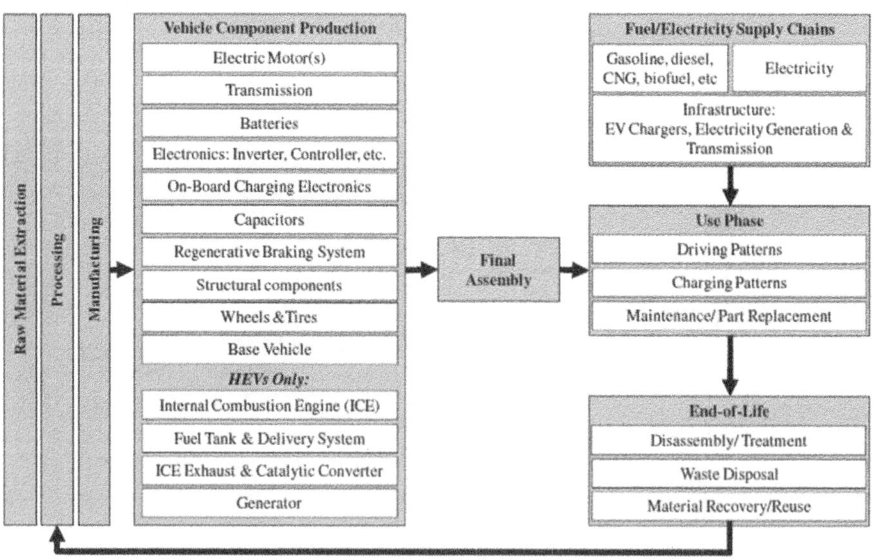

Fig. 3.10 Life cycle flowchart of hybrid electric vehicles [12]

management techniques, and the acquisition of updated online EMSs provide a benchmark. Due to a lower computing burden and no previous awareness of the entire driving cycle, the online EMSs are comparatively easy to incorporate in an actual automotive while maintaining comparable efficiency in compared to the offline EMSs [11].

Generally, hybrid vehicles are not always consumes electricity from renewable energy sources. The combustion-based power plants only supply the electric charge and supply harmful polluting gases to the atmosphere. But the emissions of the hybrid vehicles around driving surrounding such as offices, houses and its nearby places are minimized on a full hybrid drive and no emissions on the electrical drive.

Finally, the reviews give clear details for the factors affecting fuel consumption, and pollution levels of plug-in hybrid electric vehicles depend upon the energy source, charging frequency, power train construction by the manufacturer and energy source (gasoline and electricity) [12].

3.7 Present Disputes and Requirements of Hybrid Electric Vehicles

Please In the earlier decades, the cost of fuel is increased considerably, which forced the people to look for alternative efficient energy sources for vehicles. For these problems in the current scenario, the next decade promising future vehicle transport technology is given by hybrid electric vehicles [13].

Usage of renewable energy sources for charging the hybrid vehicles or battery vehicles has a lot of drawbacks due to its less power and energy density. The manufacturing cost of the hybrid vehicle is still high. The refuelling infrastructures needed huge investments to become a reality. The alternative batteries are maintained in the refuelling stations if we provide fully charged batteries to replace drained batteries to eradicate the charging time of the hybrid vehicles. On the other hand, a rapid recharge system is required for reducing the recharge time consumption. The advancement in lithium-ion batteries has less weight and rapid recharging capability, which provide a positive impact on hybrid vehicle manufacturing [12].

The vital role of the batteries and superconductors is to charge and the electrical energy through recharging, regenerative braking and discharge to drive the electric motor. Particularly, the batteries have more than one-third of vehicle weight and required maintenance in 1–2 years of the life cycle. The limited capacity of batteries needed to recharge again. The usage of superconductors in hybrid vehicles can shorten the recharging time, and the advanced power electronic controller is needed to manage the different range of discharge voltage of the energy storage.

References

1. C. Zhang, Q.X. Zeng, H. Zhu, Review on development of hybrid electric vehicles. Mech. Eng. Autom. **2**, 95 (2016)
2. A. Emadi, *Handbook of Automotive Power Electronics and Motor Drives* (CRC press, 2017)
3. J.Y. Dai Mengping, New power for vehicle-HEV system summary. Agric. Equipment Veh. Eng. **9** (2006)
4. M.R. Cuddy, K.B. Wipke, Analysis of the fuel economy benefit of drivetrain hybridization. SAE Trans. **106**, 475–485 (1997)
5. J.Y. Yong, V.K. Ramachandaramurthy, K.M. Tan, N. Mithulananthan, A review on the state-of-the-art technologies of electric vehicle, its impacts and prospects. Renew. Sustain. Energy Rev. **49**, 365–385 (2015)
6. M.A. Hannan, F.A. Azidin, A. Mohamed, Hybrid electric vehicles and their challenges: a review. Renew. Sustain. Energy Rev. **29**, 135–150 (2014)
7. Y. Chao, M. Zha, W. Wang, K. Liu, C. Xiang, Efficient energy management strategy for hybrid electric vehicles/plug-in hybrid electric vehicles: review and recent advances under intelligent transportation system. IET Intell. Transp. Syst. **14**(7), 702–711 (2020)
8. S.E. de Lucena, *A Survey on Electric and Hybrid Electric Vehicle Technology*. In Tech (2011)
9. C. Silva, M. Ross, T. Farias, Evaluation of energy consumption, emissions and cost of plug-in hybrid vehicles. Energy Convers. Manag. **50**(7), 1635–1643 (2009)
10. Ghosh, Aritra, Possibilities and challenges for the inclusion of the electric vehicle (EV) to reduce the carbon footprint in the transport sector: a review. Energies **13**(10), 2602 (2020)
11. F. Zhang, L. Wang, S. Coskun, H. Pang, Y. Cui, J. Xi, Energy management strategies for hybrid electric vehicles: review classification, comparison, and outlook. Energies **13**(13), 3352 (2020)
12. C. Samaras, K. Meisterling, Life cycle assessment of greenhouse gas emissions from plug-in hybrid vehicles: implications for policy **42**(9), 3170–3176 (2008)
13. T.R. Hawkins, O.M. Gausen, A.H. Stromman, Environmental impacts of hybrid and electric vehicles—a review. Int. J. Life Cycle Assess. **17**(8), 997–1014 (2012)

Chapter 4
Resilient Energy Storage-Based Microgrids

Kalpana Shanmugam and Pradeep Gogineni

Abstract Recent years, deploying of microgrids become key driver due to grid power un-reliability, reduction of environmental pollution (CO_2 emission) by using the renewables, demands of loads beyond the installed capacity, grid rate structure, penalties imposed due to power quality issues, incentives, etc. However, renewable energy sources will not cooperate consistently throughout the year and seasons. Changes in their availability results voltage, frequency changes causing, active, reactive power oscillations and the operation of microgrid will become unreliable. As a result, the key performances of the microgrids such as steady state, dynamic, transient response and power quality are affected. Various technical and economic challenges are introduced while integrating renewable energy sources to the utility grid. Energy storage can be a reliable asset that can bring more benefits to the microgrid. Energy storage places a critical role during the power generation fluctuations in the microgrid system due to renewables power fluctuations. It aids the microgrid to have a smooth transition from ON-grid to OFF-grid (islanded) and vice versa. In the grid connected system, it can support the real and reactive demand of the grid which can yield to shifting the energy from one time to another time. In the islanded electric system, it has the ability to operate as virtual grid for maintaining the rated voltage and frequency. Also they provide significant role in providing the spinning reserve, un-interrupted islanding, etc. In this paper, the microgrid with energy storage and renewables is analyzed in different operating conditions.

Keywords Energy storage · Grid connected · Islanded · Grid resilience · Distributed energy resources

4.1 Introduction

A microgrid (MG) is a small-scale electric grid that combines power from different distributed energy sources such as diesel, gas-operated generator sets, renewable

---

K. Shanmugam (✉) · P. Gogineni
ABB PowerGrids, Bengaluru, Karnataka 560048, India
e-mail: kalpana.s@in.abb.com

© The Author(s), under exclusive license to Springer Nature Singapore Pte Ltd. 2021
V. Kamaraj et al. (eds.), *Emerging Solutions for e-Mobility and Smart Grids*,
Springer Proceedings in Energy, https://doi.org/10.1007/978-981-16-0719-6_4

sources (PV, wind), energy storage, hydroelectric, tidal, fossil fuels, etc. MG should have the capability to control and coordinate all the connected generation assets to satisfy the requirements of load (industrial, consumer, residential). Due to the transformation toward de-centralization of the generation, we need to primarily rely on MG, which are distributed energy supply systems that are efficient, reliable, and environmentally friendly. The MG systems make social and economic development possible, changing lives by providing access to affordable, reliable electricity. The introduction of power can be a massive social and economic change for rural areas, also catalyst for social and economic development of the country. In other way, frequent power outages are major reasons for disruption in both commercial and industrial activities, resulting in economic loss and business un-success which is potentially dangerous, challenge to the country economy.

The characteristic of many microgrid components possesses more big challenges. The characteristics of many microgrid components are different from each other; hence, they bring more integration challenges as well. This paper presents the some of the technical challenges during grid integration of different renewable energy sources with possible solution by using battery energy storage system (BESS).

4.2 Microgrid System and Controls

4.2.1 Global Trends in Energy Market

The electricity production system is shifting toward the renewables. The world is adding more capacity for renewable power each year than coal, natural gas, etc. There was a shift from fossil fuels to renewables occurred in the year 2013. Also, renewable power capacity is set to expand by 50% between 2019 and 2024. The shift will continue to accelerate, and by 2030 more than four times as much renewable capacity will be The electricity production system is shifting toward the renewable energy sources. The world is giving priority to add more renewable power each year than fossil fuels. There was a shift occurred in year 2013. Also, renewable power capacity is predicted to expand by 50% within 2019–2024. The shift will continue to increase its momentum, and by 2030 more than four times renewable capacity will be installed according to the International Energy Agency. Renewables are continuously moving toward the major player seat in electricity market as utilities and grid regulators prefer renewables to replace retiring conventional generating assets. Also customers are prefer to save costs and support climate change concerns (CO_2 emissions). The price of wind and solar power continues to decline and also becoming cheaper than electricity from grid (Fig. 4.1).

In similar to renewables, the lithium-ion battery price continue to fall. Hence, BESS costs are declining. Globally energy storage capacity got doubled between 2017 and 2018 to 9 GWh [1]. A further increase to 80% is expected to happen at the end of 2020. This is due to the vast reduction in cost of batteries. The average

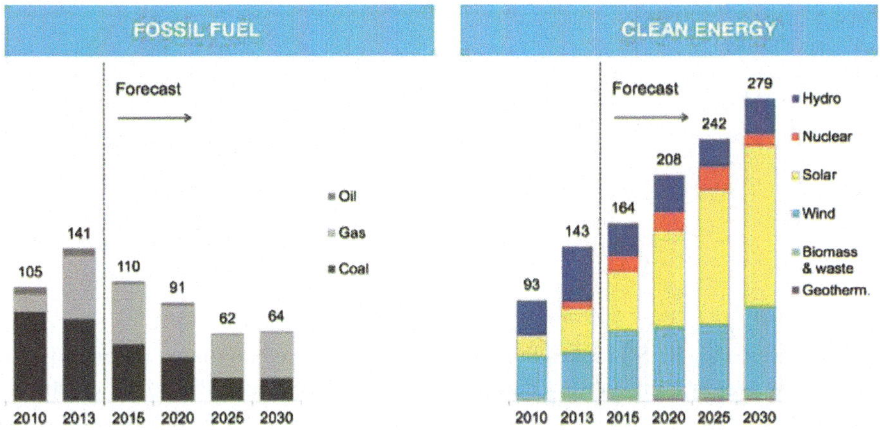

Fig. 4.1 Global trends and projections in energy market. Power generation capacity additions (GW). *Credit* Bloomberg New Energy Finance

cost of solar cells was $76.67/watt in the year 1977; it has drastically reduced to just $0.74/watt in 2013. So many customers and industry experts are moving toward renewable sources and BESS systems by forming the microgrids.

4.2.2 Renewables in Microgrid

The main advantages of renewable energy sources are the instant natural availability, less dependence on fossils fuels, reduction in cost due to technology advances, higher efficiency [2]. The positive aspects include renewables in the microgrid which can provide technical relief in line congestion, reduced losses, reduced network flows, voltage drops with reliable power in remote locations, energy security, reduction in pollution and global warming. There are also more challenging impacts due to high penetration of these renewable energy resources (RES) which include voltage fluctuations, reverse power flow, changes in power factor, injection of unwanted harmonics, frequency regulation issues, fault currents, and grounding issues and unintentional islanding [3], etc. due to unpredicted renewable power variations caused by cloud, weather and monsoon events (Fig. 4.2).

4.2.3 Battery Energy Storage System

BESS provides more flexibility to microgrid system by storing the excess energy during OFF peak period and deliver the energy to microgrid or main grid during ON peak period. Energy storage technology can vary the real power according to the

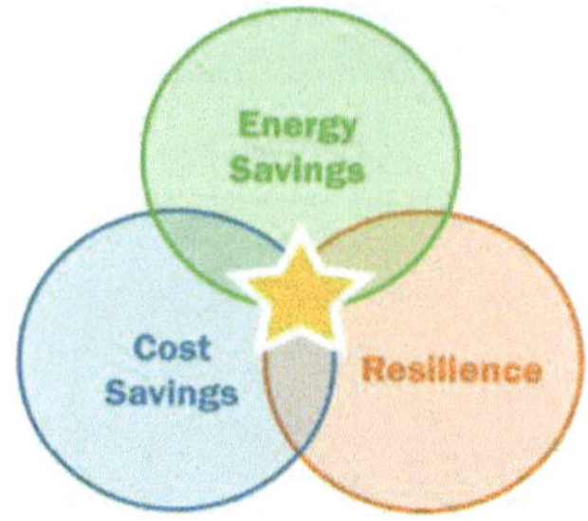

Fig. 4.2 Benefits of BESS and renewable energy resources. *Credit* US Department of Energy

changing demand. BESS makes the microgrid to be more economic, efficient [4]. It improves power system stability and power quality. BESS uses the power electronic converter (inverter) which converts the DC voltage of battery to the AC voltage at required voltage amplitude, phase angle required for the main grid or microgrid [5]. It can operate in both virtual generator mode (VGM) and grid supporting mode (GSM). The d-axis and q-axis current parameters are controlled to control the real and reactive power supplied by BESS. BESS significantly supports microgrid during frequency recovery phase [6].

4.2.4 BESS Power Capabilities

BESS is suitable for most of the applications due to its unique characteristics of working in both as the power generation source and also as a load. It has the capability to respond fast and precisely for the variation in the control signal (V, F). Also have the capability of providing reactive power services (both supply and consumption), with relatively high accuracy and efficiency. The reactive power according to the active power availability depends on the PQ capability curve of BESS. The active power depends on the availability of reactive power, PQ curve shape, frequency, parameters of frequency regulation. The priority of functionalities that can deliver the good profit to a BESS operator is sensitive to the system market prices and rules, the power need of the local distribution grid. More analyses and intelligence functional algorithms need to be added to define the optimum priority of tasks.

4.2.5 Droop Control in Islanded and Grid Connected System

The droop control is a very popular method for controlling power converters in parallel with in microgrids generating resources [7]. In conventional droop control, frequency relatively decreases with the reduction of the active power supplied. Voltage droop control is a popular methodology for controlling power electronic converters operating in parallel with in microgrids [7]. The active power sharing between power converters is obtained in precise way by using the droop characteristics. The imprecise load sharing using droop control can produce either deep discharge or overcharge in the BESS. Efforts are required to match the settings of Q-V droop control with the line impedances. Otherwise, the Q-V droop control gives errors in the distribution of reactive power. Dedicated communication systems are used for eliminating/reducing the error on reactive power sharing.

4.2.6 Microgrid Architecture

The basic concept in traditional power system is its centralized control architecture with unidirectional energy flow from generating sources to load centers. The centralized secondary control consists of one central secondary controller which collects the information from all the primary controllers in the systems and sends back the calculated set points to the primary controllers using dedicated communication channels. The centralized control approach requires point-to-point communication, which adds complexity, less reliability as there could be chances for a single-point failure. Alternatively, distributed secondary control method has more attractive features; hence, they became more popular. In the distributed control architecture, secondary control consists of equivalent number of primary controllers. Each secondary controller gets information through the dedicated communication network from all the other secondary controllers, and the set points are evaluated smartly based upon the gathered information (Fig. 4.3).

In this paper, the microgrid controllers are operating in distributed control system architecture. All the generating resource controllers share the information themselves

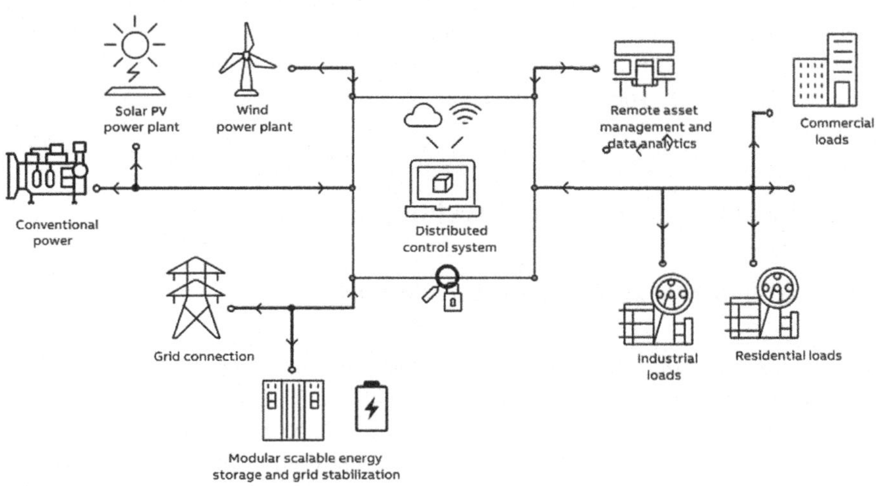

Fig. 4.3 Benefits of BESS and renewable energy resources

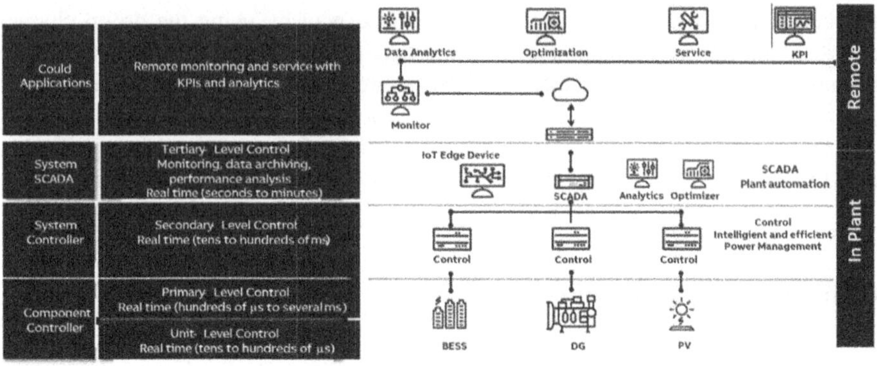

Fig. 4.4 Benefits of BESS and renewable energy resources

and respond based upon the operating conditions of the microgrid system. So any of the controllers are failed or lost from the communication, then the remaining controllers control the microgrid system without any issues.

4.2.7 Proposed Coordinated Primary and Secondary Control

Our proposed MG system is implemented with different control layers as depicted in Fig. 4.4. The generating resources (PV, BESS, DG) are having its own primary-level controllers. The microgrid secondary-level controllers control the primary controllers to operate the microgrid system in economical, optimized and desirable operating scenarios. The system level has the data reporting and analytics operations. It also sends the user-level set points to the secondary-level controllers. Some of the key point indices and graphs are externalized to the cloud. So that the user can be viewed these values from anywhere to know about the microgrid system behavior.

4.3 The Developed Control Technique

4.3.1 Renewable Integration Challenges

In the olden days, due to the absence of generating units connected, distribution systems are called passive circuits. But, now with the RES, the situation is reverse. Depending upon the generation capacity, RES can be integrated to the grid at the transmission level (large capacity) or at the distribution level (small capacity). Majority of the RES systems are connected at LV distribution level. The grid integration of RES depends on a numerous parameters like the share of RE power, size and location of network in which it is connected, energy conversion technology, the effect on system inertia, droop, power quality, system protection, etc. [8, 9]. According to IEEE Std. 519, Point of Common Coupling (PCC) should be a point which is accessible to both utility and customer for direct measurement [10]. The variations in RE result in voltage variations at the PCC. The strength of the grid characterized by its short-circuit ratio (SCR) is defined in [11]. A grid can be titled as STRONG with respect to the RE integration if SCR is above 10. If SCR value of a bus is less than 5, it is usually not recommended to connect RE to that system [12]. In conventional distribution system the bus voltage variation is maintained within an acceptable range. But with RES in the system, voltage will go beyond the upper limit; another issue is the unbalanced voltage. The unbalanced distribution systems can produce problems for RES connected to it, as the resulting unbalanced currents in the generators can undergo frequent shutdowns due to overheating caused. The power quality problem can defined as any unacceptable deviations in magnitude, frequency, or waveform shape of the voltage and current that can cause failure or malfunctioning of customer equipment [13]. In case of PV systems, power quality problems arise due to variations in solar radiation, cloud shadow, power electronic modules such as inverter and filters due to their nonlinear mode of operation. The power quality issues like over voltage, short-time voltage fluctuations, frequency deviations, unbalance, etc. arises due to fluctuating nature of energy resource (PV, wind). The other type of power quality issues like harmonics, resonance, inrush currents are caused due to the power electronic converter interface with the power system.

4.3.2 Possible Solution Approaches for Challenges

If SCR value of a bus is less than 5, it is usually not recommended to connect RE to that system [12]. PQ problems can be mitigated by two methods. One from the customer/load end another from the grid/utility end. The first is known as load conditioning, to ensure that the connected equipment is less sensitive to power supply/disturbances, allowing the normal operation even under voltage disturbances. The second one is by installing line conditioning equipment which can suppress or offset the power quality disturbances.

4.3.3 Operational Control of Proposed System

Distributed control architecture with dedicated peer-to-peer communication between the secondary controllers is proposed in this paper which provides more flexibility and simplicity, scalability and modularity to the microgrid. Control logics can be developed and deployed across the secondary controllers at any time. This allows enhancements that must be made to meet new market requirements and/or the addition of new resources/controllers/field devices, to be made efficiently and accurately,

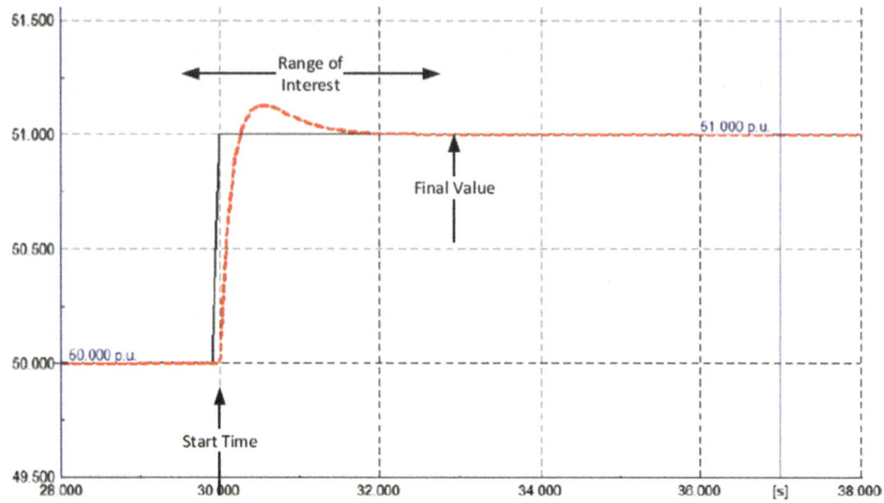

Fig. 4.5 Compensation for the native response at the secondary level

easing and simplifying updates and new equipment integration into the existing MG system architecture.

The setpoints are calculated in the secondary controllers based upon the primary controllers' actual values. The actual microgrid system should follow the single-order response (PT1) for attaining the required setpoint command. The secondary controller continuously calculates the setpoint such that the primary controller responds in smoother way. All the secondary controllers are communicating with each other through the dedicated peer-to-peer communication network in real world, through the MG table in the simulation environment. The compensation for the native response of each asset in MG has been done in the secondary so that any dissimilar generating assets (generator, BESS) will have the same (compensated) response irrespective of their native response, make, type, etc. Hence, we will be unrestricted, we can work with any type of generation assets, load sharing and frequency control now identical irrespective of the generation asset type. The compensation for the native response has been done for P, Q, V, F. BESS can be operated in VGM and GSM modes (Fig. 4.5).

4.3.4 Droop-Based Secondary Control

We have used the droop-based secondary control due to its widely accepted benefits in the scientific community. The droop control strategy is that it changes the output of active and reactive power to control the frequency and output voltage, so that microgrid system can work on stabilized voltage, frequency setpoints in islanded operation mode (Fig. 4.6).

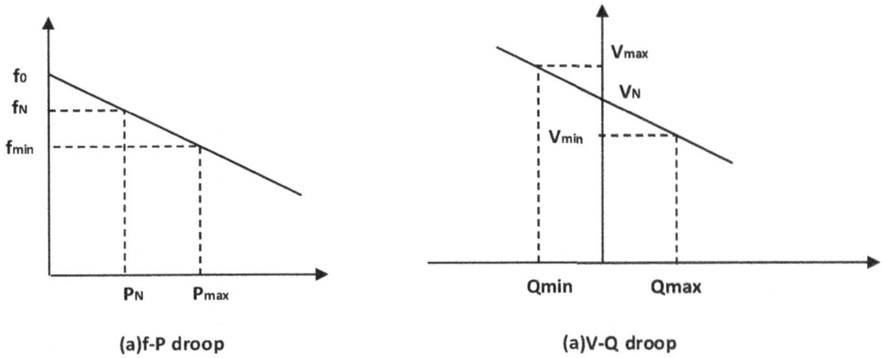

Fig. 4.6 Drooping curve: **a** F-P droop and **b** V-Q droop

4.4 Proposed Simulation Model and Results

Simulations are performed in DIgSILENT power factory which is conventionally used software for power system modelling. It has the ability to conduct steady state, time domain (quasi-dynamic, RMS and EMT) RMS modeling simulations. Globally, these simulations are used for the grid analysis/study. These simulations are also referred to as (dynamic) stability simulations; the advantage of using RMS simulations is that the simulation speed can be increased/improved significantly. We have used DIgSILENT Simulation Language (DSL) for model definition and composite frames for defining the MG resources. DIgSILENT also provides the flexibility to either simulate with a fixed time step or a variable time step (Fig. 4.7).

The primary and secondary controllers of all MG sources are modeled as composite frames in the DIgSILENT. The secondary control is linked by using dll. At any time the user can switch ON/OFF the assets by switching off their corresponding frame defined.

4.4.1 Design Specifications

See Table 4.1.

4.4.2 Peak Lopping

This is the ability of the microgrid to ensure that, at the point of coupling with grid, the import and export values remain between predefined limits defined by the grid operator. PV power to be curtailed only if diesel generators and BESS are at their minimums and we are exporting into the power grid up to the maximum limit

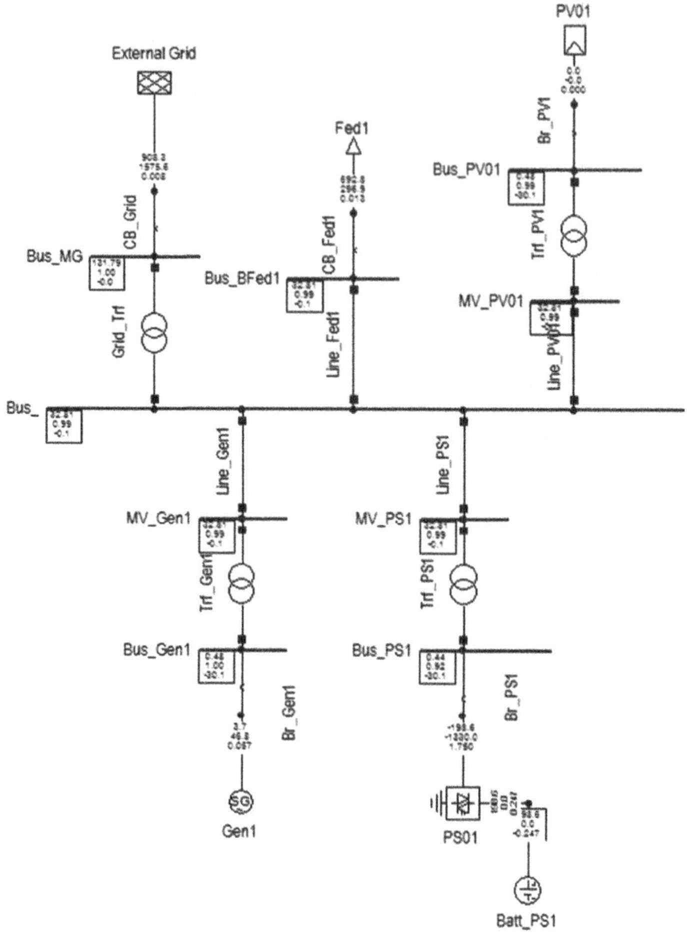

Fig. 4.7 Proposed simulation model of the MG system with DERs

Table 4.1 Ratings of the microgrid assets

Microgrid asset	Power system rating	Remarks
Diesel generator 1	1250 kVA	Min loading 300 kW Max loading 700 kW
Photovoltaic	1250 kVA plant	
BESS	1000 kVA/500 kWh (Power converter/battery)	Charge limit −200 kW Discharge limit 400 kW
Feeder load	Feeder-2000 kVA	
Network/grid	2000 kVA	Import limit 200 kW Export limit 300 kW

allows. Prioritize imports from the grid up to the allowed limit rather than using the diesel generators as grid electricity is cheaper while respecting minimum allowable loading on the generators. Prefer discharging BESS (after maximum network import limit has been reached) to using diesel to cover demand. This function lives on three controllers: PV, BESS, and generator (Fig. 4.8).

The MG system is running with DG, PV, BESS, and grid connection. At 95 s, the load power 1061 kW is supported by 600 kW of PV, 300 kW of DG, and 180 kW of grid. The DG is running at the minimum loading, and MG is importing the grid power up to its limit. At 120 s, the load is increased to 1350 kW. BESS is supporting the increased load since the grid already reached its import limit. At 180 s, the load power further increased to 1640 kW. Once the BESS reaches its discharge power limit, the additional required load power is supported by DG. At 250 s, the PV power is increased to 760 kW. So the DG supporting power is decreased and maintaining at its minimum loading of 300 kW. At 320 s, the load power is further decreased to 1220 kW, and the BESS supporting power is decreased from its discharge limit. At 420 s, the load is decreased to 836 kW. The BESS power is reached up to its charge limit of −200 kW. The grid importing power is reduced to 0 kW. The microgrid secondary controllers are operating such that grid power is maintained within the allowable limits.

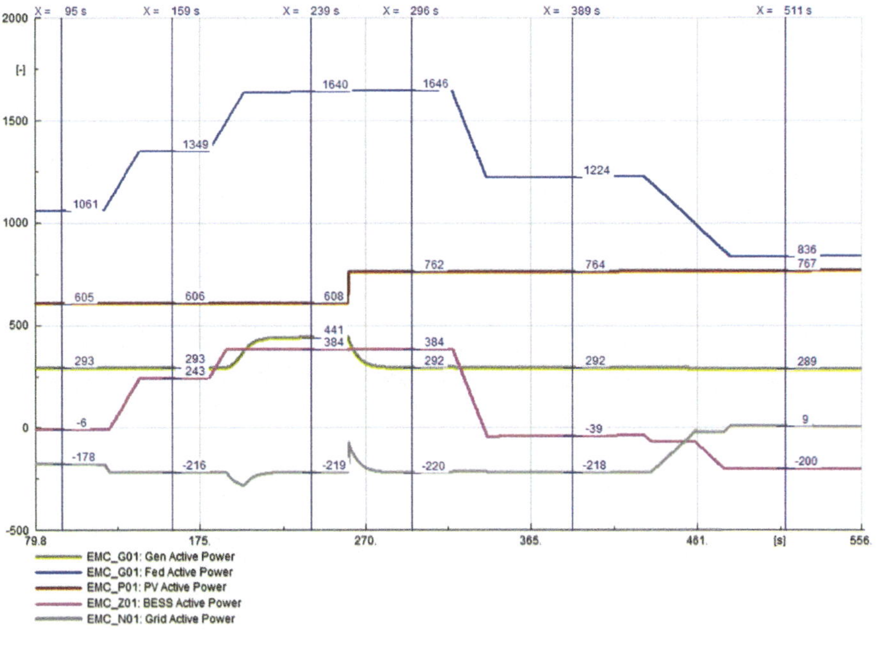

Fig. 4.8 Peak lopping behavior of microgrid control system

4.4.3 Smoothing

The output from renewable is always a varying one without any warnings. This impacts the diesel generators operation and their efficiency. The energy stored in BESS can be used to counteract the variations in renewable energy sources. The renewable source and BESS could work in concert to create a total power output that is less variable. Although the BESS could be used to fill in all the valleys and cutoff all the peaks, creating a dead flat composite output, the microgrid can usually sustain power injection with some variation. The grid operators also limit the variation of the power at the PCC point. Renewable power variation should not affect the grid power variation. It should be smoother way as per agreed limits of grid operator (Fig. 4.9).

The MG system is running with PV, BESS, and grid connection. The BESS charging limit is decreased to 100 kW due to its state of charge. Initially, the load power is at 600 kW. The PV is supplying 1000 kW. The grid is exporting 300 kW up to its export limit. BESS is absorbing the 100 kW. At 200 s, the PV power suddenly reduced to 680 kW. The sudden 320 kW power deficit is supported by BESS initially after that the BESS supporting power is reduced drastically as per the allowable ramp rate provided by grid operator. So the grid exported power is reduced 37 kW as per the ramp rate. At 250 s, the PV power is increased to 900 kW. The extra available power is absorbed by the BESS to ensure the smooth variation of the grid power at PCC. The grid exported power reached to 240 kW as per the allowable ramp rate.

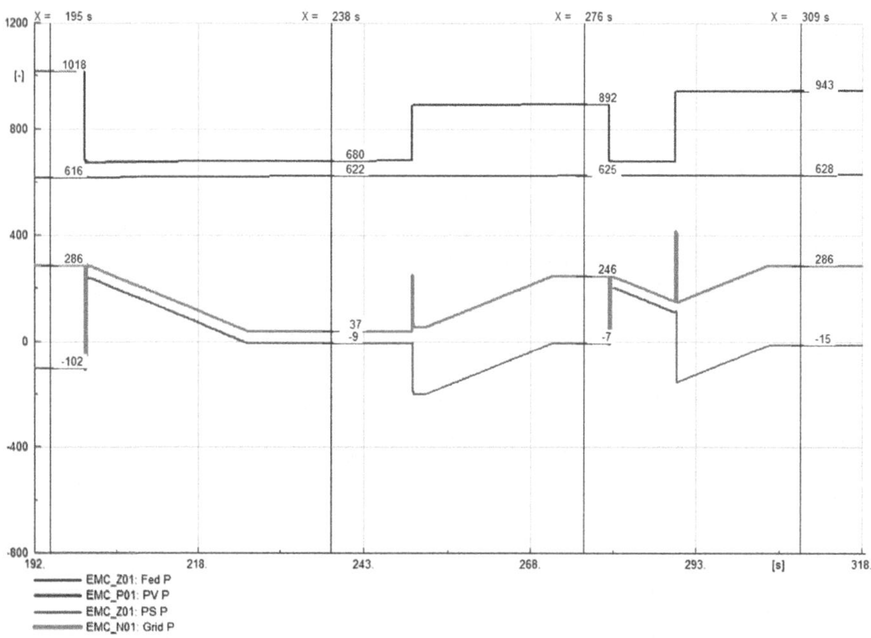

Fig. 4.9 BESS smoothing behavior of microgrid control system

Two more sudden PV cloud events are created at 280 and 290 s. The grid power is varying smoothly at the PCC. The sudden variations are absorbed by BESS.

4.4.4 Transitions

In case the microgrid can operate both on-grid and off-grid, it is necessary to manage the transition from on-grid to off-grid coordinating the microgrid assets so that the grid breaker can be opened with minimum power flow across it so to avoid unnecessary transients. This function receives the unloading command and reduce the power flow (both active and reactive) across the grid breaker within the specified limits. While it receives the grid connection command, the voltage phasors on the two sides of the grid breaker are synchronized so that the breaker can be closed and parallel the microgrid with the main grid without overcurrents, trips and equipment damage. MG control system sends the synchronization relay increase/decrease voltage (frequency) signals into voltage (frequency) setpoints for the microgrid grid forming generators and BESS. Once the frequency and voltage signals are satisfied as per synchronization process, the grid circuit breaker is closed afterward (Fig. 4.10).

The MG system is running with PV, BESS, and grid connection. Initially, the load power is at 600 kW. The PV is supplying 400 kW due its availability. The grid is

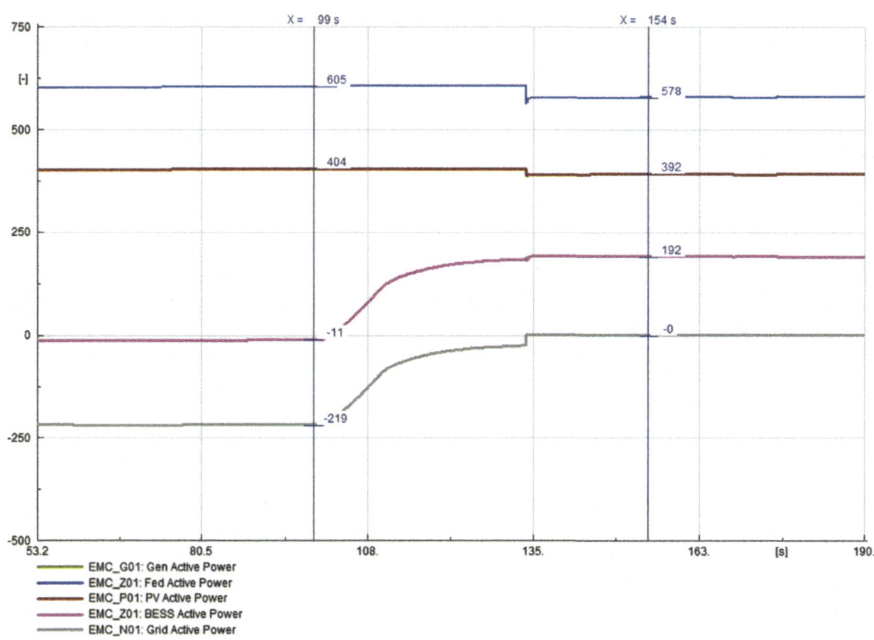

Fig. 4.10 Grid transition to islanded behavior

importing 200 kW up to its import limit. BESS is absorbing the 10 kW. At 100 s, the planned islanded mode operation is triggered. The microgrid control system responds to its event and reduces the grid import power to zero. Once the grid power falls nearer to zero, then the grid circuit breaker open command issued.

4.5 Conclusion

More penetration of RESs is expected in MG systems as they are more economical, eco-friendly, pollution-free, less CO_2 emission, etc. In this paper, the challenges in the integration of microgrids like control and protection are discussed. RES technological improvement, advancement in energy storage systems can help the new microgrid system to become economically viable to consumer. Integration of renewable sources to the grid is beneficial and efficient thing in all aspects.

This work highlights how BESS provided the flexibility to the microgrid using the distributed control architecture. The change of power setpoints to BESS as per agreed power limits at PCC is showcased with the simulations. Also how BESS counteracts for fluctuations in renewables so that the power will be reliable and stable. The simulation results have been discussed including seamless transition from grid to island connected. In this paper, we have presented performance of the system through software-in-the-loop test results. Also a proven industrial solution 21 and [14] has been verified and validated using DIgSILENT PowerFactory simulation model. Also ABB's microgrid projects are included in the reference sections.

References

1. https://www.energy.gov
2. B. Khan, P. Singh, Selecting a meta-heuristic technique for smart micro-grid optimization problem: a comprehensive analysis. IEEE Access Open Access J. **2017**(5), 13951–13977 (2017)
3. J.Y. Huang, C.W. Jiang, R. Xu, A review on distributed energy resources and microgrid. Renew. Sustain. Energy Rev. **12**(9), 2472–2483 (2008)
4. L.-J. Qin, W.-T. Yang, Micro-grid converter droop control strategy and simulation. Adv. Electr. Comput. Eng. ISBN: 978-1-61804-279-8
5. G. Wang, G. Konstantinou, C.D. Townsend, J. Pou, S. Vazquez, G.D. Demetriades, V.G. Agelidis, A review of power electronics for grid connection of utility-scale battery energy storage systems, in *IEEE Transactions on Sustainable Energy*, vol. 7, no. 4 (2016), pp. 1778–1790
6. C. Wanichrojanarat, P. Wirasanti, Control strategy for seamless transition of microgrid using battery energy storage system, in *IEEE Transactions on Power Systems*
7. ABB Reference projects, http://www.abb.com/microgrids
8. G. Lalor, A. Mullane, M. O'Malley, Frequency control and wind turbine technologies. IEEE Trans. Power Sys. **20**(4), 1905–1913 (2005)
9. J. Morren, S.W.H. de Haan, W.L. Kling, Wind turbines emulating inertia and supporting primary frequency control. IEEE Trans. Power Sys. **21**(1), 433–434 (2006)
10. IEEE Std. 519-2014, IEEE Recommended Practice and Requirements for Harmonic Control in Electric Power Systems

11. P. Kundur, *Power System Stability and Control* (McGraw Hill, 1994)
12. V. Gevorgian, S. Booth, Review of PREPA technical requirements for interconnecting wind and solar generation, in NREL Technical Report, NREL/TP-5D00-57089, November 2013. http://www.nrel.gov/docs/fy14osti/57089.pdf
13. IEEE Std. 1159-2009, IEEE Recommended Practice for Monitoring Electric Power Quality
14. R. Hidalgo-Leon, P. Jacome-Ruiz, C. Sanchez-zurita, J. Wu, Y. Munoz-Jadan, Roles, challenges, and Approaches of droop control methods for Microgrids, in *IEEE PES Innovative SmartGrid Technologies Latin America*, December 2017
15. M.T. Yeshalem, B. Khan, Microgrid integration, in *Intechopen.78634*. DOI: 10.5772, Published: Novemeber 5 (2018)
16. Dr. K.V. Vidyanandan, B. Kamath, Grid integration of renewables: challenges and solutions, in *Emerging Energy Scenario in India–Issues, Challenges and Way Forward* (2018)
17. R. Walling, A. Ellis, S. Gonzalez, Implementation of voltage and frequency ride-through requirements in distributed energy resources interconnection standards, in Sandia National Laboratories Technical Report 2014-3122, California, April 2014
18. https://www.renewableenergyworld.com, https://press.trendforce.com/node/view/3333.html
19. ABB Reference projects, https://new.abb.com/news/detail/2290/the-energy-example-of-robben-island-mandelas-prison
20. ABB Reference projects, https://new.abb.com/news/detail/13052/abb-state-of-the-art-microgrid-to-power-african-hub-of-the-international-committee-of-the-red-cross
21. Digsilent GmbH, DIgSILENT PowerFactory Application Example, Battery Energy Storing Systems BESS. Gomaringen, Germany
22. ABB Solutions. https://new.abb.com/distributed-energy-microgrids
23. ABB control Solutions. https://new.abb.com/distributed-energy-microgrids/our-offering/e-mesh/control
24. Large Scale Grid Integration of Renewable Energy Sources - Way Forward, Central Electricity Authority, India, Nov. 2013
25. BP Statistical Review of World Energy, June 2019

Chapter 5
Electric Traction Over Head Equipment Protection Using Intelligent Electronic Device

K. N. Dinesh Babu and Salman Khan

Abstract Protection relays used in traction system application utilize impedance protection having parallelogram characteristics with three zones of protection comprising independent R, X values with forward, reverse and independent time settings. These relays are configured to protect the 25 kV single phase 50 Hz traction transmission line which is fed from traction substation with intermediate neutral sections and insulating overlap (IOL). IOL provides further complications when the locomotive moves in this section, where the pantograph located on top of the locomotive may connect the phases of two traction substations, resulting in a wrong phase coupling scenario. Such scenarios should be detected, and the circuit has to be isolated by distance protection relay which senses the fault in the second quadrant of the R-X plane. In addition to this scenario, when regenerative breaking is applied, the moment of inertia will result in the traction motors acting as generators which will result in pumping of current into the traction line resulting in the impedance measurement in second quadrant of the R-X plane. Moreover, the distance relay algorithm is challenged by overloading condition when a locomotive enters into a section. In addition, very high resistive fault would not have sufficient current to drive the impedance relay is the additional challenge. With the availability of latest relay testing equipment, testing of these characteristics has been automated, and the process of mimicking such characteristics and testing it with smart relay testing equipment is discussed in this paper.

Keywords Distance protection parallelogram characteristics · Wrong phase coupling · Traction system · Neutral section · Panto flashover · Delta current protection · Regenerative braking · Intelligent electronic device

K. N. Dinesh Babu (✉) · S. Khan
Megger, Chennai, India
e-mail: dineshbabu.nagalingam@megger.com

S. Khan
e-mail: salman.khan@megger.com

© The Author(s), under exclusive license to Springer Nature Singapore Pte Ltd. 2021
V. Kamaraj et al. (eds.), *Emerging Solutions for e-Mobility and Smart Grids*,
Springer Proceedings in Energy, https://doi.org/10.1007/978-981-16-0719-6_5

5.1 Introduction

Indian Railways is one of the major consumers of electric power and the electrified sections which enforce the need for protection of the equipment. It is the fourth largest railway network in the world after USA, Russia and China running about 20,000 trains carrying more than 2.5 crore passengers and 2.8 million tonnes of freight every day. The network has been expanding since 1853 when railways were introduced in India. The expansion has reached nearly 65,000 km of permanent way out of which 60% of freight traffic and 48% of passenger traffic is electrified [1]. The electrified routes are fed from traction substation and switching substations which are installed with traction transformer fed from the state and central electricity boards. These traction transformers convert three phase of various voltages like 132 kV, 110 kV or 66 kV to single phase 25 kV 50 Hz supply which runs from one station to another station with capacitor banks installed for power factor control. In between two stations, the continuity of the overhead line has to be maintained, and at the same time, the lines have to be switched from one station to another without breaking the contact. This is achieved by the introduction of neutral section and IOL. These equipments, discussed so far comprising the traction transformer, overhead equipment (OHE) and the capacitor banks, need to be protected to ensure proper disturbance free operation of the system. In this paper, a detailed discussion on the testing procedure is taken up, and the required fundamentals [2] on the protection philosophy are discussed to understand the testing process.

Section 5.2 talks about the distance protection feature provided to the OHE. Section 5.3 details the procedure to draw the parallelogram characteristics in relay test management software (RTMS) which is used to control smart Megger relay tester (SMRT) equipment. Section 5.3 talks about wrong phase coupling (WPC) protection. Section 5.4 talks about delta current followed by panto flashover protection in Sect. 5.5. Section 5.7 talks about auto reclosure followed by conclusion.

5.2 Distance Protection for Overhead Traction Line

Figure 5.1 shows the parallelogram characteristics of a distance protection relay in R-X plane, which is used to protect the OHE. The relay is equipped with three zones of protection in single phase mode to protect against phase to earth faults. Zone 1 is the primary zone of protection, and zone 2, zone 3 act as backup protection in the forward and reverse direction, respectively. The angle denoted as reach characteristics angle (RCA) is in the tune of 70°. This value is based on the resistance (R) and the reactance (X) of the OHE utilized, which is normally hard drawn grooved copper contact wire of 107 mm^2 size [3].

The distance protection settings denoted as RF, RB, XF and XB will be calculated independently for each zone as per the guidelines provided by Indian Railways [5]. The procedure to draw this characteristic is discussed in the following section.

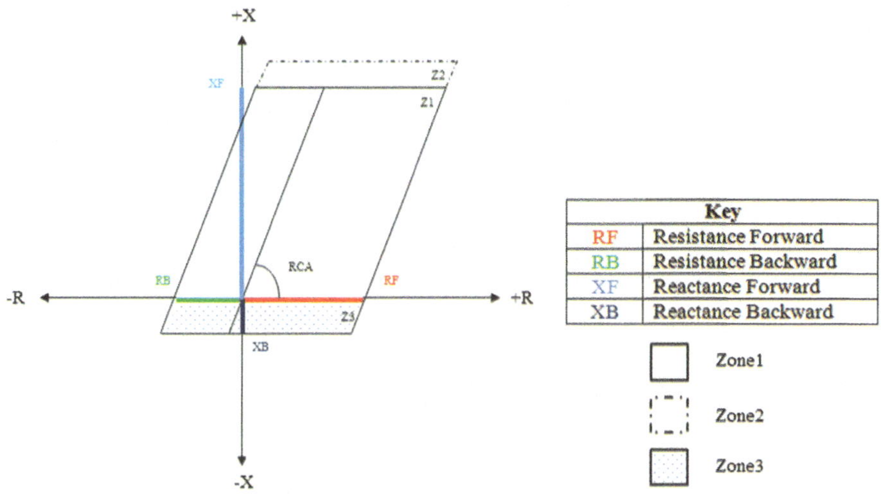

Fig. 5.1 Parallelogram impedance characteristics [4]

Figure 5.2 shows the quadrilateral characteristics available in RTMS software with which majority of the characteristics of a distance protection relay can be plotted. These characteristics are used to draw the parallelogram characteristics of the OHE protection relay. Various settings such as the current transformer ratio, potential transformer ratio, primary/secondary settings, timing for each zone, load encroachment,

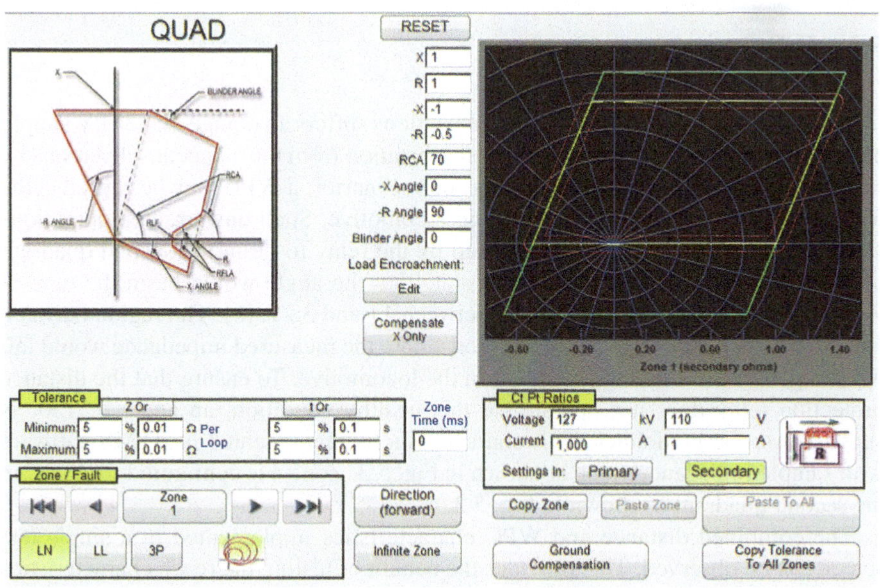

Fig. 5.2 Quad characteristics of RTMS

Table 5.1 RTMS parallelogram characteristics sample settings of Fig. 5.2

Railway terminology	RTMS terminology	Zone 1	Zone 2	Zone 3
XF	X	1	1.2	1
RF	R	1	1	1
XB	−X	−1	−1	−0.5
RB	−R	−0.5	−0.5	−0.5
RCA	RCA	70	70	70
	−X angle	0	0	0
	−R angle	90	90	90
	Blinder angle	0	0	0

zero sequence compensation and tolerance for testing results can be configured as in Fig. 5.2.

The R and X values of the OHE can be entered as shown in Table 5.1 in order to achieve the parallelogram characteristics for each zone. The grey areas shown do not have any impact in the drawing as these settings are applicable for non-directional configuration only. In this manner, the faults in the OHE can be identified and tripped; however, in certain scenarios, a phase to phase fault occurs which cannot be detected by the distance protection relay since it is a phase to ground protection. In order to protect against such faults, a protection called as WPC is discussed in the next section.

5.3 WPC

In scenarios where one of the traction substations suffers an outage, the 25 kV supply may be extended up to the failed traction substation from the adjacent substations by closing the bridging interrupter. During such scenarios, a WPC may be caused at the overlaps by the pantograph of a passing locomotive. Such unwarranted operations would result in the impedance measured by the relay to lie in the second quadrant of the R-X plane. Railway guidelines state that the angle would be in the tune of 90° to 150° with an impedance lying between 11 and 38 Ω [6]. The region between 150° to 180° is the reverse feeding region where the measured impedance would fall when regenerative braking is applied in the locomotive. To ensure that the distance protection relay does not operate for this healthy condition, an angle of 150° is suggested in the guidelines. This characteristic is implemented in RTMS software with sample impedance value as shown in Fig. 5.3. Zone 4 is configured for WPC in the second quadrant as shown in Fig. 5.3.

The combined distance and WPC characteristics implemented in a single test screen can be observed. This provides the benefit of testing the reach characteristics and timing in a single test. In this manner, the impedance characteristics can be used

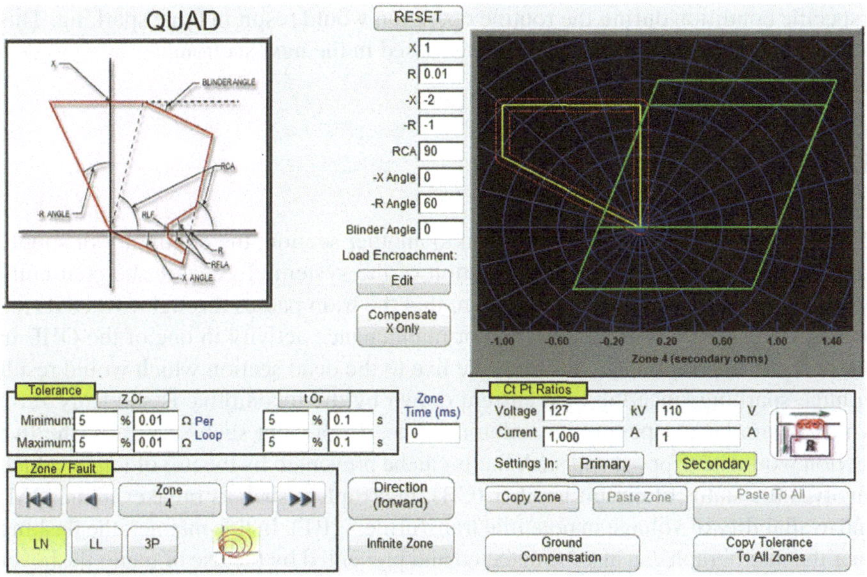

Fig. 5.3 WPC characteristics in RTMS (sample settings)

to identify the faults related to OHE protection based on the principle of measure current and voltage. In general, the current always increases, and the voltage dips during a fault which is utilized in these impedance-based protections; however, in case of a high resistive fault, the dip in voltage and increase in current may not be sufficient. In order to handle such high resistive faults, an alternate detection process is required which is discussed in the next section

5.4 Delta *I*

A high resistive fault may not be detected by the distance protection relay, and to ensure such faults are identified and isolated, the vectorial difference between the base current and the fault current is identified. If the vectorial difference is greater than the settings, the relay will identify it as a Delta I (ΔI) fault. In addition, the load current has harmonics due to the movement of locomotive, whereas fault current will have more fundamental component which is also utilized in ΔI detection. This harmonic principle is based on the concept that second harmonic will be predominantly present during charging of a transformer (locomotive transformer) and third harmonic component will be present during load current variation.

With the use of configuration and testing of distance, WPC and ΔI, the OHE can be protected against phase to ground, phase to phase and high resistive faults; however, a specific condition during the routine operation would result in huge sparking. This condition and the ways to protect are discussed in the next section.

5.5 Panto Flash Over

When the train passes from one section to another section, the continuity of supply has to be maintained for proper operation of the system. To ensure the continuity of the supply, the use of IOL will ensure that the train passes through two OHE fed from different TSS. In case of a fault or maintenance activity in one of the OHE in one section, the train may pass from the live to the dead section which would result in huge sparking since the load current drawn by the locomotive is suddenly zero, which is similar to opening an isolator online. To prevent such scenarios, the live section would be tripped, and such faults can be prevented by the use of logics which involves the status of circuit breaker (CB), interrupters such as breaker main (BM) and availability of voltage in potential transformers (PT). In this manner, the flashing near the pantograph can also be detected and prevented by the use of panto flashover protection. In addition to this protection, many faults are transient in nature, and to ensure automatic clearance of the transient faults, the concept of auto reclosure is adapted which is discussed in the following section.

5.6 Auto Reclosure

Several fault in the OHE is transient in nature, and these may be caused due to several factors such as deposit of carbon on the OHE emitted by diesel locomotives, deposit of salt in coastal regions, lightning, falling of branches from nearby trees, surface flashover due to pollution in atmosphere. Such faults would be cleared automatically within few seconds when the line is discharged for a short duration and then reclosed.

The relay performs this function by the use of a protection control function called as auto reclosure. In this function, the relay detects a fault and issues a trip command to the circuit breaker. The breaker opens, and the line is dead for a short duration. The short duration for which the breaker is opened is called as the dead time. On completion of the dead time, the relay issues a close command to the circuit breaker. On successful closure of the circuit breaker, the relay monitors the parameters to ensure there is no consecutive fault for a period of time. This period of time is called as the reclaim time. In case a fault occurs during the reclaim time, it is assumed that the fault is not transient and hence the line has to be tripped permanently which is termed as lockout.

5.7 Conclusion

A transmission line is protected based on the comparison of the impedance set in the relay with the actual impedance measured by the relay. The impedance measured is a ratio of current and the voltage measured which is compared with the impedance settings of the relay. The impedance settings which are set in the relay remain a constant; however, traction system OHE protection is one of the complex protection system due to the movement of locomotive. The movement of locomotive will result in a variable impedance which acts as a challenge for the protection system. To ensure proper operation of the protection system, several protection functions have been implemented as per the railway standard.

On the other hand, the advancement in numerical relays with inbuilt and user programmable logics enhances the complexity for the user during testing of these features. To minimize the challenges and provide a simplified testing guideline, this paper shares the procedure that may be adopted to better understand the philosophy involved in testing.

References

1. http://www.irieen.indianrailways.gov.in/view_section.jsp?lang=0&id=0,294,366. Last accessed 2020/09/15
2. Ministry of Indian Railways, Traction distribution by Indian Railways institute of electrical engineering, 98 (2010)
3. Indian Railways Institute of electrical engineering, Over Head Equipments, 6 (2010)
4. Indian Railways, Centre for advanced maintenance technology, Power Supply installation, PhaseII-Module No.STC-TrD-05
5. ACTM & RDSO Guideline, Instruction No.TI/IN/0017 (July2008) & L.No.TI/PSI/PROTCT/STATIC/07 (23/04/2007)
6. Spec No:TI/SPC/PSI/testkit/1080 (12/2008)

Chapter 6
Differential Evolution Based Design Optimization of Flywheel with Different Materials

V. Ramya, R. Ramaprabha, and M. Balaji

Abstract A flywheel is one of the most important renewable sources of power transmission and storage. It has a wide range of applications such as light rail power, hybrid electric vehicles, industrial pulse power and is employed for power quality improvement. It is a rotating structure capable of transmitting power through the rotation of the shaft. A flywheel made from the filaments of composite material provides high specific energy. This paper discusses differential evolution-based optimization procedure to determine the parameters of the flywheel to produce the required energy storage with minimum mass. The procedure is carried out considering different materials to select the best material for the required energy storage. The results of the differential evolution algorithm are compared with particle swarm optimization, and the closeness of the result validates the proposed approach.

Keywords Flywheel energy storage · Optimization · Differential evolution · Particle swarm optimization

6.1 Introduction

Flywheel is one of the most important sources of renewable energy which causes no environmental pollution and has very high energy density. It is a rotating mass that stores kinetic energy. Though the manufacturing cost is high, it gives a very long life span and thus makes a perfect choice for power systems. Due to seasonal and peak load, power consumption reaches the maximum values in a particular time zone of the day which can cause power quality problems such as voltage sag and harmonic distortion. In such case, the flywheel can act as auxiliary source. There has been a wider research into the design and analysis of flywheel energy storage system. Optimization procedures have been proposed to arrive at an appropriate flywheel design to produce the required energy with minimum mass. A design optimization procedure to evolve optimal flywheel design has been discussed in [1]. The design

V. Ramya · R. Ramaprabha (✉) · M. Balaji
Department of EEE, Sri Sivasubramaniya Nadar College of Engineering, Kalavakkam, Chennai 603110, India
e-mail: ramaprabhar@ssn.edu.in

© The Author(s), under exclusive license to Springer Nature Singapore Pte Ltd. 2021

V. Kamaraj et al. (eds.), *Emerging Solutions for e-Mobility and Smart Grids*,
Springer Proceedings in Energy, https://doi.org/10.1007/978-981-16-0719-6_6

procedure is formulated to maximize the kinetic energy of the flywheel with mass and stress as constraints. A design procedure for optimizing the flywheel design for vehicles has been proposed in [2]. In [3], two approaches to the design of reinforced composite flywheels are presented to maximize the kinetic energy of the flywheel. Performance evaluation of different flywheel geometry for automotive applications has been researched upon in [4]. The optimization of flywheel materials and geometrical structure has been carried out using optimization techniques [1, 5]. The above literatures highlight the need for design optimization procedure and choice for proper material to arrive at a proper design of flywheel to suit the application requirements. In this prelude, this paper focuses on differential evolution-based optimization procedure to determine the optimal parameters of the flywheel that produces the required energy with minimum mass.

6.2 Energy and Inertia

Energy stored in a flywheel depends on moment of inertia (I) about its spin axis, which depends on mass and radius of the flywheel.

The total stored energy in a flywheel is given by

$$\Delta E = I\omega^2 \Delta C = mr^2\omega^2 \Delta C \tag{6.1}$$

where ω is angular velocity and ΔC is coefficient of fluctuation of speed.

The coefficient of fluctuation of speed is assumed to be one so that the equation now becomes

$$\Delta E = mr^2\omega^2 \tag{6.2}$$

The above equation can be used to determine the mass

$$m = \Delta_E / r^2\omega^2 \tag{6.3}$$

The maximum speed with which a flywheel can rotate depends on hoop stress [2] it develops, which is given as

$$\text{Hoopstress} == \sigma = \rho r^2\omega^2 \tag{6.4}$$

'ρ' is the density of the given material, and 'σ' is the hoop stress of the given material.

The above equation can be rewritten as

$$\omega^2 \leq \left[\frac{\sigma}{\rho r^2}\right] \tag{6.5}$$

Table 6.1 Material properties

Material	Allowable stress $[\sigma]$ (MPa)	Density $[\rho]$ (kg/m^3)
Gray cast iron (GCI 25)	220	7340
Maraging steel	900	8000
Carbon fiber composite (40% epoxy)	750	1550
E glass fiber (40% epoxy)	250	1900

Equation (6.5) is rewritten as

$$\rho r^2 \omega^2 - [\sigma] \leq 0 \tag{6.6}$$

From the above equation, it is evident that the radius, the material density and operating speed influence the hoop stress. The safe operation of flywheel is ensured when the hoop stress developed is within the allowable stress.

This work engages the use of differential evolution-based optimization that produces the required energy with minimum mass considering angular velocity and radius as variables. The optimization problem is carried out for different materials to identify the suitable material that produces the required energy with minimum mass and ensures that the flywheel is operated within the stress [1, 6, 7]. The material properties and their stress values are tabulated in Table 6.1, [1, 6, 7].

6.3 Differential Evolution

In this work, differential evolution algorithm [8] has been applied to solve the optimization problem. Differential algorithm has been preferred to determine optimal solutions owing to its merits like ease of implementation, small number of control parameters and exhibits better convergence. The algorithm employs arithmetic operators and genetic operators to evolve the final solution [9, 10].

In differential evolution algorithm, the target vectors $X_i(t)$ are the members of the initial population. The mutant vector is produced by performing mutation operation. The trial vector is generated by combining mutant vector with target vector. The members of the next generation are selected by evaluating the fitness of the trial vector and comparing it with the target vector. The step-by-step procedure for differential evolution algorithm is discussed below.

Step 1: Initialization

The first step involves initializing the parameters of the algorithm which include number of members in the population, constraints, the mutation factor (F), the crossover rate (CR) and maximum number of generations.

Step 2: Generate the members of the current population

The initial members of population corresponding to each variable in optimization are generated randomly.

Step 3: Fitness evaluation of the current population

The mass of the each member of the current population is evaluated.

Step 4: Mutation operation

The mutation operation creates a donor vector $V_i(t)$. The mutation process for jth component of each vector is expressed by equation.

$$V_{i,j}(t+1) = X_{r1,j}(t) + F \cdot (X_{r2,j}(t) - X_{r3,j}(t)) \tag{6.7}$$

In the above equation, the members X_{r1}, X_{r2} and X_{r3} are generated randomly from the current population, and F is the mutation factor. This mutation operation strategy is labeled as DE/rand/1 [8].

Step 5: Crossover operation

The crossover operation generates the trial vector $Ui(t)$ as per the equation illustrated below

$$\begin{aligned} u_{i,j}(t) &= v_{i,j}(t) \ \text{ if } \ (\text{rand}(0,1)) < \text{CR} \\ &= x_{i,j}(t) \ \text{ if } \ (\text{rand}(0,1)) > \text{CR} \end{aligned} \tag{6.8}$$

In this operation, if the randomly generated value is less than crossover rate, then the donor vector will compete in the selection process.

Step 6: Selection operation

The fitness of the trial vector is evaluated, and upon exhibiting better fitness, the trial vector participates in the next generation. Once a new population is generated, the process of mutation, crossover and selection is repeated until the maximum number of generations is reached.

6.4 Results and Discussion

The algorithm is coded in MATLAB environment [11], and the parameters of the algorithm are enumerated in Table 6.2. The flowchart of the DE algorithm is given in Fig. 6.1.

Table 6.2 Differential evolution algorithm parameters

Population size, NP	40
Crossover constant, CR	0.7
Scaling factor for mutation, F	0.8
Maximum generations, iter_max	100

Fig. 6.1 Flowchart of differential evolution algorithm

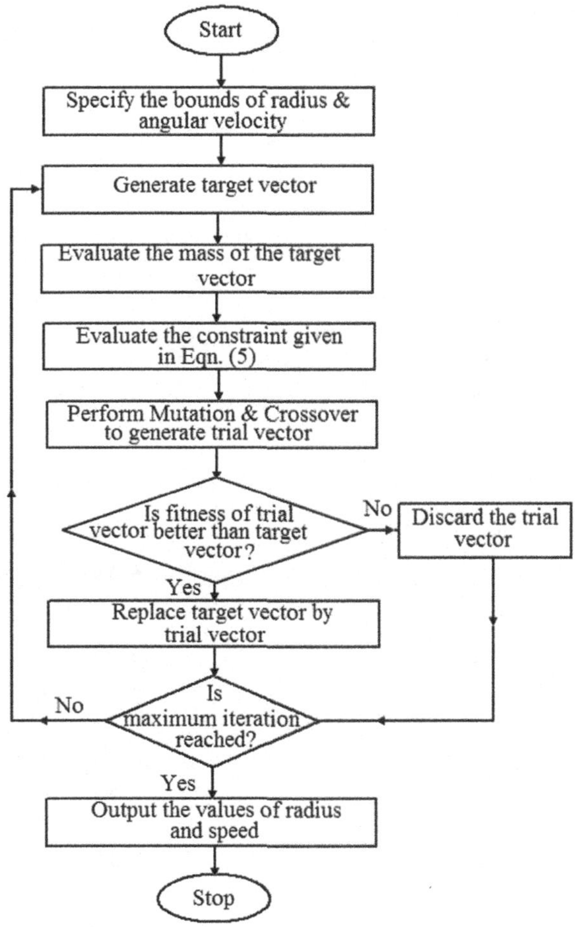

To validate the proposed approach, the results of differential evolution algorithm are compared with particle swarm optimization-based approach, and the results are tabulated in Table 6.3. The closeness of the results validates the application of differential evolution algorithm to determine the material that produces the required energy with minimum mass. The result indicates that the carbon fiber composite material produces the required energy with lesser mass. The convergence characteristics of the differential evolution algorithm and particle swarm optimization algorithm are depicted in Fig. 6.2. Figure 6.2 illustrates the fact that the particle swarm optimization algorithm has better convergence in comparison with differential evolution algorithm.

The optimization is carried out considering varying the required energy and thus determining the radius and angular velocity. The results of optimization considering carbon fiber composite with different energy levels are tabulated in Table 6.4, while

Table 6.3 Results of optimization

Material	Energy storage [ΔE] (J)	Mass [m] (kg)		Radius [r] (m)		Angular velocity [ω] (rad/s)	
		DE	PSO	DE	PSO	DE	PSO
Gray cast iron (GCI 25)	100	0.00347	0.00347	1.69	1.66	100	101.55
Maraging steel	100	0.00087	0.00087	1.43	1.44	234	233.88
Carbon fiber composite (40% epoxy)	100	0.0002	0.0002	1.68	1.6	416	428
E glass fiber (40% epoxy)	100	0.0007	0.00073	2	1.98	184	186

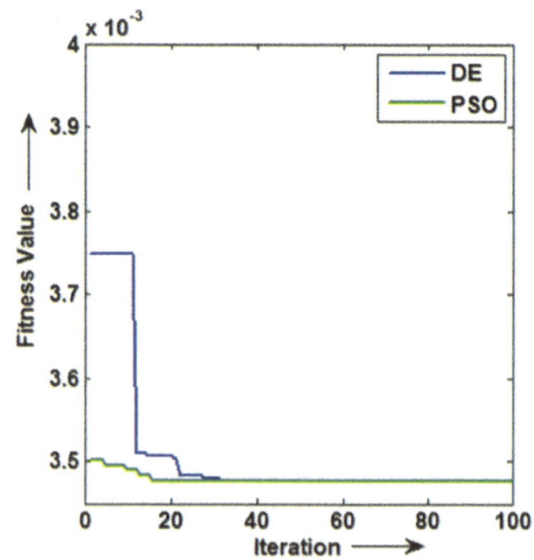

Fig. 6.2 Convergence characteristics of differential evolution and PSO algorithm

Table 6.4 Results of optimization for different energy levels with carbon fiber composite material

Material	Energy storage [ΔE] (J)	Mass [m] (kg)	Radius [r] (m)	Angular velocity [ω] (rad/s)
Carbon fiber composite (40% epoxy)	200	0.0004	1.67	417.78
	300	0.0006	1.68	414
	400	0.0008	1.60	436
	500	0.001	1.63	428

Table 6.5 Results of optimization for different energy levels with E glass fiber material

Material	Energy storage [ΔE] (J)	Mass [m] (kg)	Radius [r] (m)	Angular velocity [ω] (rad/s)
E glass fiber (40% epoxy)	200	0.0014	1.26	293
	300	0.002	1.20	307
	400	0.0029	1.26	293
	500	0.0036	1.28	288

the results with E glass fiber are given Table 6.5. From the tables, it is evident that the carbon fiber composite material produces the required energy with lesser mass.

6.5 Conclusion

A differential evolution-based optimization procedure to determine the parameters of the flywheel has been discussed in this paper. The optimization algorithm has been formulated to determine the radius and angular velocity for the specified energy level with minimum mass. The optimization has been carried out considering gray cast iron, maraging steel, carbon fiber composite and E glass fiber. The results reveal that the carbon fiber composite material yields the required energy with minimum mass. The results of the algorithm have been validated using particle swarm optimization approach. The closeness of the results validates the differential evolution-based approach in determining the optimal parameters of the flywheel energy storage.

References

1. F. Baronti, S. Vazquez, M. Chow, Modeling, control and integration of energy storage systems in E-transportation and smart grid. IEEE Trans. Industr. Electron. **65**(8), 6548–6551 (2018)
2. M. SenthilKumar, Yogesh Kumar, Introduction to magnetic materials using genetic algorithm. Int. J. Mech. Ind. Technol. **5**(2), 6–9 (2017)
3. K. Dems, J. Turant, Two approaches to optimal design of composite flywheel. Eng. Optim. **41**(04), 351–363 (2009)

4. M. Hedlund, J. Lundin, J. de Santiago, J. Abrahamsson, H. Bernhoff, Flywheel energy storage for automotive applications. Energies **8**, 10636–10663 (2015)
5. Y. Li, C. Zhu, L. Wu, Y. Zheng, Multi-objective optimal design of high-speed surface-mounted permanent magnet synchronous motor for magnetically levitated flywheel energy storage system. IEEE Trans. Magn. **55**(7) (2019)
6. A. Nayar, *The Metals Databook* (Tata McGraw Hill, New Delhi, 2009)
7. K.K. Chawla, *Composite Materials Science and Engineering*, 2nd edn. (Springer Verlag, NewYork, 2001)
8. R. Storn, K. Price, Differential evolution—a simple and efficient heuristic for global optimization over continuous spaces. J. Global Optim. Arch. **11**(4), 341–359 (1997)
9. R. Ramaprabha, B.L. Mathur, Enhanced maximum power point tracking using differential evolution algorithm for partially shaded solar photo-voltaic array. Int. J. Electr. Eng. Embed. Syst. **2**(2), 159–164 (2010)
10. T. Kamf, High speed flywheel design using advanced composite materials, Uppsala University, Disciplinary Domain of Science and Technology, Technology, Department of Engineering Sciences, Electricity (2012)
11. www.mathworks.com

Chapter 7
Comparison of Radial Flux PMSM and Axial Flux PMSM for Hybrid Electric Tracked Vehicles

S. T. Vigneshwar and N. C. Lenin

Abstract Permanent magnet synchronous motors are increasingly becoming predominant with the upscale competitiveness of high-power permanent magnets. Permanent magnets are comparatively more efficient because the field excitation losses are extinguished that leads to significant rotor loss minimization. Thus, the motor efficiency is greatly surpassed, and higher power density is achieved. In addition to this, they have better torque, lower acoustic noise and mechanical vibration. In this paper, a PMSM is proposed for an axial flux permanent magnet synchronous motor after emphasizing its advantages over the radial flux permanent magnet synchronous motor. The motor with the focus of being applied for an enormous 110 kW application is designed and compared with radial flux permanent magnet synchronous motor.

Keywords PMSM · Axial flux permanent magnet synchronous motor (AF-PMSM) · Power density · Efficiency · Synchronous reluctance motors · Radial flux permanent magnet synchronous motor (RF-PMSM)

7.1 Introduction

The permanent magnet synchronous motor seems to be the best option for off-road traction motor vehicle. PMSMs can randomly achieve three times their nominal torque at low speeds and more than two times their nominal rotating speed with better efficiency, while the drop in torque remains appropriate. Typical properties incorporate high torque capability at lower speeds, high power density, wide operating speed range, high efficiency and high reliability at sensible cost. PMSM can be a better substitute for diesel engine in terms of economy. With the several advancements, researchers have proposed many configurations of PMSM. One among those works,

S. T. Vigneshwar
School of Engineering, VIT Chennai, Chennai 600127, India

N. C. Lenin (✉)
Electric Vehicles-Incubation, Testing and Research Centre, VIT Chennai, Chennai, India
e-mail: lenin.nc@vit.ac.in

© The Author(s), under exclusive license to Springer Nature Singapore Pte Ltd. 2021

V. Kamaraj et al. (eds.), *Emerging Solutions for e-Mobility and Smart Grids*,
Springer Proceedings in Energy, https://doi.org/10.1007/978-981-16-0719-6_7

de Almeida et al. [1] has proposed a paper on axial flux PMSM, showing better efficiency features. Rallabandi et al. [2] proposed a novel AF-PMSM topology, obtained by combining a yokeless central stator type NS arrangement and observed very high flux concentration. M. Aydin et al. [3] proposed a novel new coreless spoke type AF-IPM motor and compared with conventional coreless AFPM motor after which he observed better torque features in the novel coreless spoke type. Bo Zhang et al. [4] presented three permanent magnet topologies and compared them from the obtained results he concluded that the AFM with segmented armature torus structure can be a better substitute for conventional radial flux machine with the advantage of reduced axial length. A. A. Pop and M. Radulescu [5] evaluated axial flux and radial flux PM machines and observed that the AF-PMSM provided higher torque to weight ratio, better heat removal, adjustable air gap along with several other advantages over its rival. Qurban et al. [6] proposed and presented a new single-phase flux switching axial flux permanent magnet motor. After analysing its characteristics features, it is observed that the motor possesses symmetrical back EMF and reduced cogging torque.

7.2 Modelling of Radial Flux PMSM

The stator and the rotor made up of laminated silicon steel. The rotor field windings are replaced by permanent magnets in order to increase efficiency. The radial flux PMSM is illustrated in Fig. 7.1.

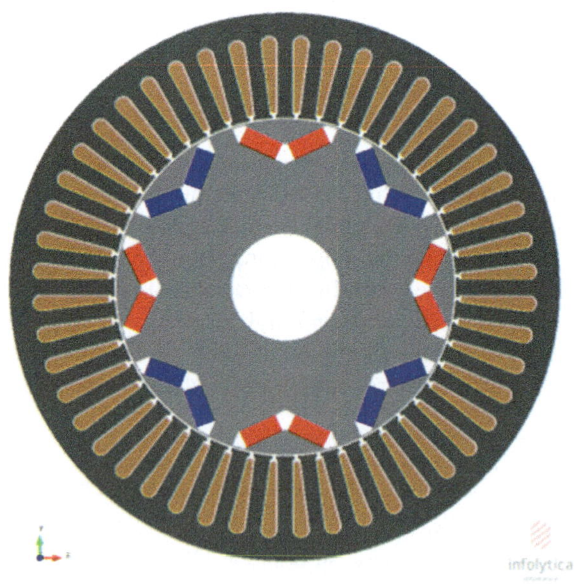

Fig. 7.1 Structure of radial flux PMSM

Table 7.1 Dimension details of radial flux PMSM

Parameter	Values
Stator outer diameter (mm)	400
Stator inner diameter (mm)	239
Air gap length (mm)	0.5
Machine length (mm)	120
Number of stator slots	48
Stator tooth width (mm)	10.7
Stator slot depth (mm)	61
Rotor outer diameter (mm)	238
Shaft diameter (mm)	77
Number of rotor pole	8
PM width (mm)	25
PM thickness (mm)	12
PM orientation (°)	25
Number of turns/coil	9
AC voltage (Vrms)	800
SWG number	15
Number of strands	8
Slot fill factor (%)	46.1
Winding connection type	STAR

The magnets are buried inside the rotor core to protect against demagnetization fields and mechanical stress. The flux is concentrated, and hence, high air gap flux density is achieved which is in proportion with torque. The constructional details of the RF-PMSM are given in Table 7.1. The outer diameter of the machine is 400 mm, while the air gap is kept 0.5 mm. The application of interior PM to the motor made it more feasible for the motor to function flawlessly even with minimal air gap.

The result parameters are discussed in Table 7.2, and most of the important results are discussed. At the rated speed of 1050 rpm, the motor obtains 1045 Nm of torque

Table 7.2 Results of radial flux PMSM

Parameters	Values					
Speed (rpm)	1050	1500	2000	2500	3000	3500
Input current (A)	230	193	166	145	137	130
Current density (A/mm^2)	17	13.7	12	10.6	9.8	9
Total loss (kW)	9.1	8.1	7.1	6.2	5.8	6.3
Efficiency (%)	90.9	91.9	92.9	93.8	94.2	93.7
Torque (Nm)	1045	829	667	512	412	357

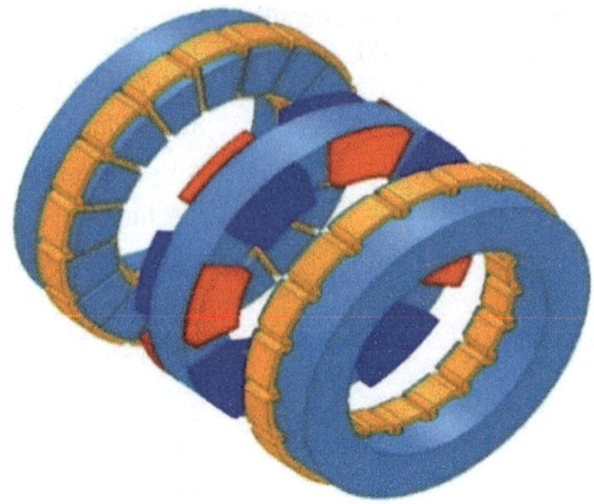

Fig. 7.2 Structure of axial flux PMSM

with the consumption of 230 A input current. The efficiency obtained is 90.9 which is predominant.

7.3 Modelling of Axial Flux PMSM

The AFM stator is made up of soft magnetic material. The stator is wound with double layer winding which resembles a toroidal structure for short end connection. Unlike the radial flux PMSM, the permanent magnets are placed in the axial direction.

The availability of air gap on both the sides of the motor improves torque density. The structure of the proposed axial flux PMSM is illustrated in Fig. 7.2. The AF-PMSM comprises two rotors sandwiching double-sided rotor on either side.

The dimension details of the axial flux PMSM are given in Table 7.3. The outer diameter of the machine is kept 350 mm without the inclusion of the casing. The air gap is kept 1.4 mm on each side. The total length of the machine is 106.8 mm.

The results obtained after the simulation are tabulated in Table 7.4. At the rated speed of 1050 rpm, the machine torque obtained is 1303 Nm with the consumption of 149 A of input current.

7.4 Comparison of RF-PMSM and AF-PMSM

Both the radial flux permanent magnet synchronous motor and axial flux synchronous motor results are compared together. The simulation results are based on the electromagnetic analysis with the help of Ansys simulation tool.

Figure 7.3 illustrates the comparison graph of current versus speed of both the machines, and it is observed that at 1050 rpm, the RF-PMSM draws 35% more current

Table 7.3 Dimension details of axial flux PMSM

Parameters	Values
Machine OD (mm)	350
Machine ID (mm)	150
Air gap length (mm)—for one sided	1.4
Stator stack length (mm)—for one sided	42
Rotor core stack length (mm)	10
Magnet axial thickness (mm)	5
Magnet embrace	0.85
Magnet radian length	80
Total rotor stack length (mm)—for one sided	15
Total machine length (mm)	106.8
Number of stator slots	24
Number of rotor poles	20
Stator and rotor material	M19-29G
Magnet material	NdFeB-1.4T
AC voltage (Vrms)	800
Number of phases	3
Number of conductors per slot	20
Slot fill factor (%)	47
Wire diameter (mm)	3.459

Table 7.4 Results of axial flux PMSM

Parameter	Values					
Speed (rpm)	1050	1500	2000	2500	3000	3600
Average input current (A)	149	101	75	58	51	44
Current density (A/mm^2)	13.7	10.5	8.9	7.7	6.8	6
Total loss (kW)	7.77	6.98	6.1	5.5	5.23	5.7
Efficiency (%)	92.23	93.02	93.9	94.5	94.77	94.3
Machine torque (Nm)	1303	855	643	516	413	366

than AF-PMSM. The current graph starts with a spike followed by near constant line. The current drawn dips in drastically once the speed reaches 3500 rpm.

The speed versus current density comparison is depicted in Fig. 7.4, and the inference indicates that the results are in line with the current versus speed parameter. The graph initiates with a little rise at the rated speed and then begins to settle post 1500 rpm. At the speed of 1050 rpm, the RF-PMSM records for 19% more current density than AF-PMSM.

The loss versus speed comparison graph is portrayed in Fig. 7.5, and the results indicate that at the speed of 1050 rpm the AF-PMSM accounts for minimal loss in

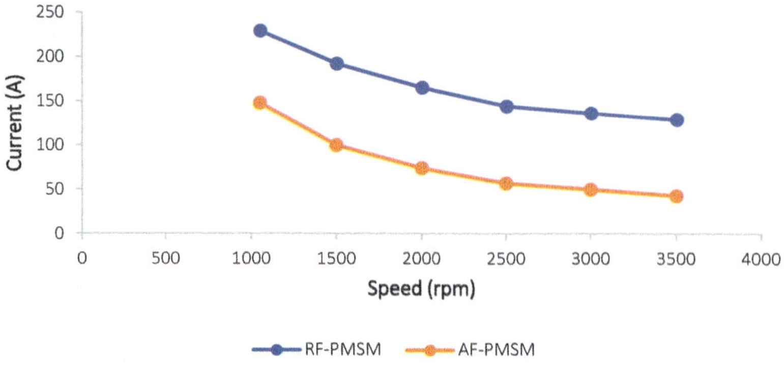

Fig. 7.3 Current versus speed comparison

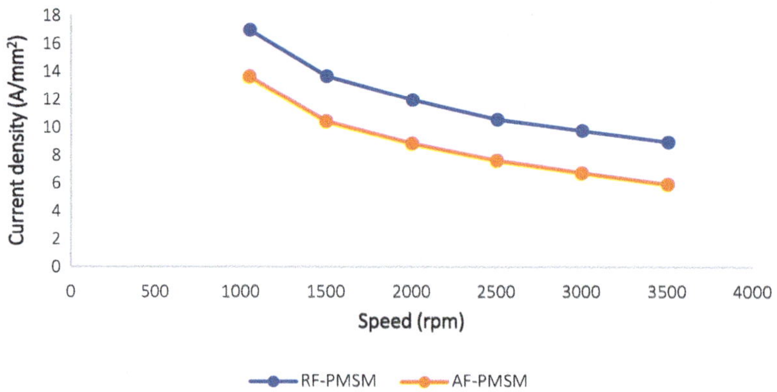

Fig. 7.4 Current density versus speed comparison

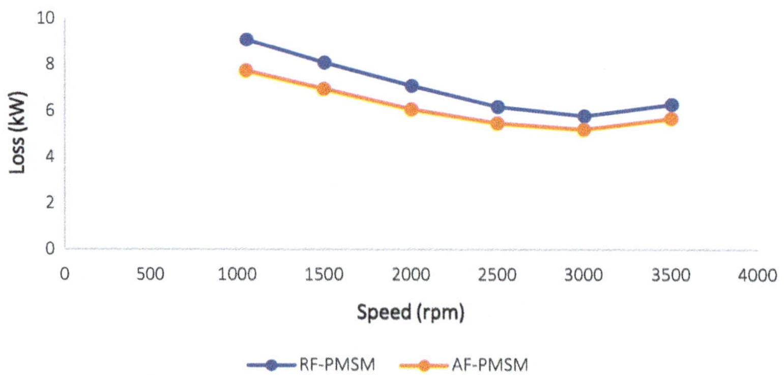

Fig. 7.5 Loss versus speed comparison

contrast with the RF-PMSM. The difference obtained is 14.6% more loss value than AF-PMSM. The loss graph is curvy starting with a spike at 1050 rpm and again loops in post 3000 rpm.

The efficiency comparison between the AF-PMSM and RF-PMSM is depicted in Fig. 7.6, and the results clearly indicate that at the speed of 1050 rpm the AF-PMSM outperforms RF-PMSM by 1.46% in the efficiency department as well. The efficiency which is inversely proportional to loss follows the expected trend. The efficiency boosts up once the speed sets in.

The torque comparison results are depicted in Fig. 7.7, and the results represent that the AF-PMSM comprises better torque characteristics by 24.6% at 1050 rpm in contrast with RF-PMSM. The initial spike in the AF-PMSM is followed by a sequential torque rating with the radial flux PMSM post 1500 rpm.

The instantaneous current and torque values are being incorporated in the upcoming two figures; the figures are being exported directly from the FEA-based tool [7].

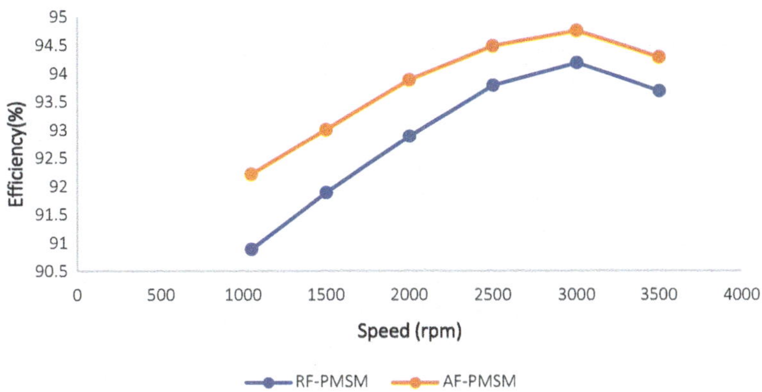

Fig. 7.6 Efficiency versus speed comparison

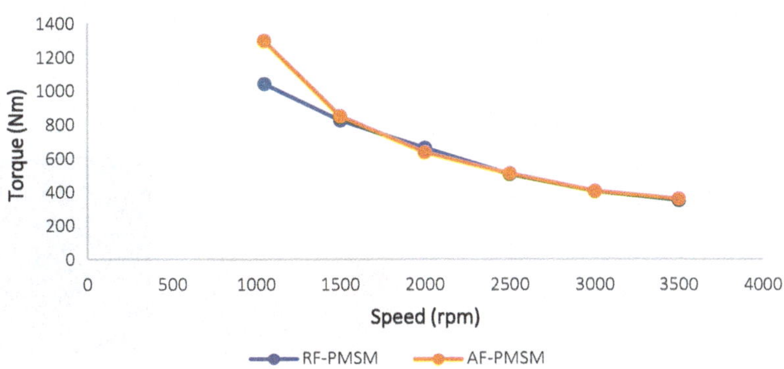

Fig. 7.7 Torque versus speed comparison

The instantaneous current values for various speeds are depicted in Fig. 7.8.

The instantaneous torque values for various speeds are illustrated in Fig. 7.9.

The comparison of specific power capabilities between AF-PMSM and RF-PMSM is illustrated in Fig. 7.10, and it is observed that the AF-PMSM has an improved marginal difference of 97% specific power capabilities with its rival.

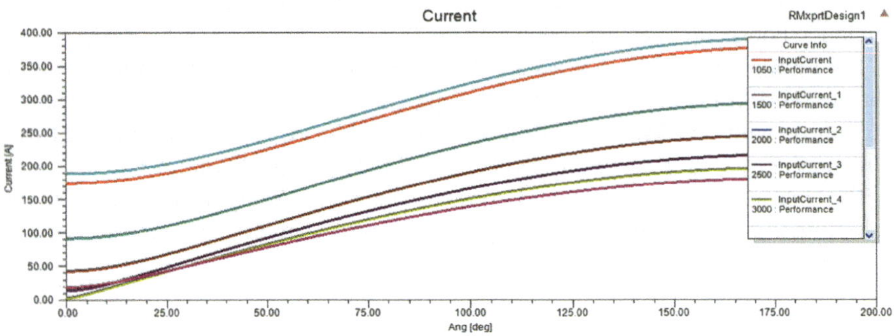

Fig. 7.8 Instantaneous current at different speeds

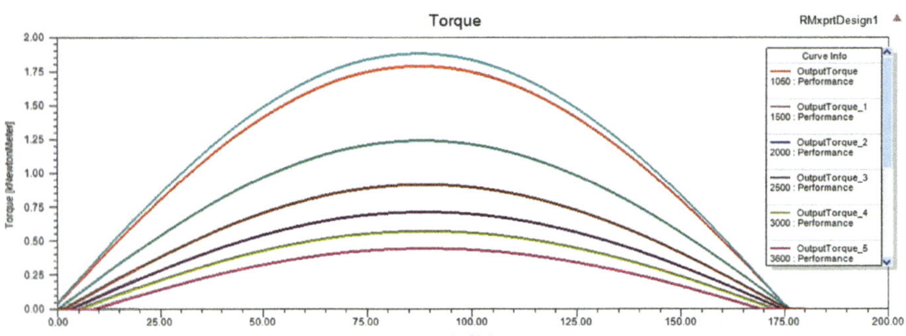

Fig. 7.9 Instantaneous torque at different speeds

Fig. 7.10 Specific power versus speed comparison

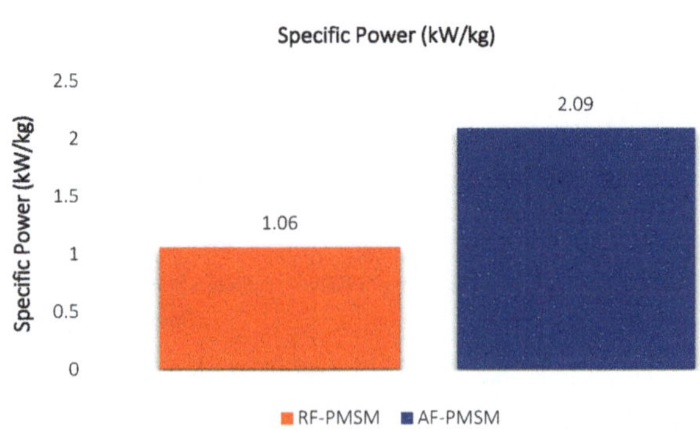

The machine length comparison of the rivals is graphed in Fig. 7.11, and the results indicate that RF-PMSM is 11% lengthier than the AF-PMSM. Thus, the precise structure adds up another advantage to AF-PMSM.

The weight comparison between the AF-PMSM and RF-PMSM is depicted in Fig. 7.12, and the results are distinctly supporting the AF-PMSM which is around 51% less weighted than its rival.

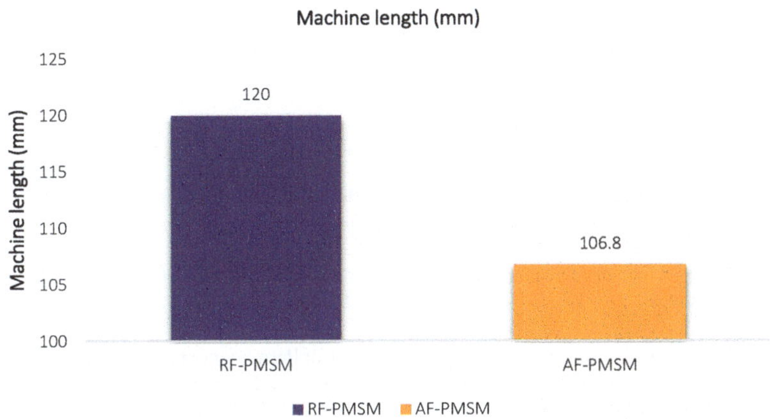

Fig. 7.11 Machine length comparison

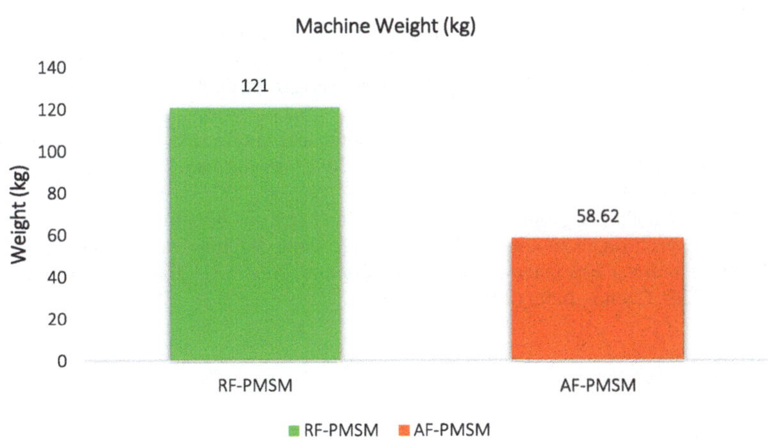

Fig. 7.12 Machine weight comparison

7.5 Conclusion

As far as the electromagnetic analysis is concerned, the proposed axial flux PMSM outperforms the conventional radial flux PMSM in all the criteria and fits in perfect for the 110-kW application. With the add on benefit of lightweighted structure, the motor finds wider range of application in the fields of automobile, military and commercial applications. The motor catches the catch word smaller size more power and tops the table comfortably with its rivals.

References

1. A.T. de Almeida, F.J.T.E. Ferreira, G. Baoming, Beyond induction motors—technology trends to move up efficiency, in *IEEE Transactions on Industrial Applications*, vol. 50, no. 3 (2014), pp. 2103–2114
2. V. Rallabandi, N. Taran, D.M. Ionel, I.G. Boldea, An ultra-high specific torque PM axial flux type motor with flux focusing and modulation, in *2017 IEEE Energy Conversion Congress and Explosion (ECCE)* (2017), pp. 1–5
3. M. Aydin, M. Gulec, A New Coreless Axial Flux Interior Permanent Magnet Synchronous Motor With Sinusoidal Rotor Segments, in IEEE Transactions on Magnetics, vol. 52, no. 7, pp. 1-4, July (2016), Art no. 8105204. https://doi.org/10.1109/TMAG.2016.2522950
4. Bo. Zhang, T. Epskamp, M. Doppelbauer, M. Gregor, A comparison of the transverse, axial and radial flux PM synchronous motors for electric vehicle, 2014 IEEE International Electric Vehicle Conference (IEVC), Florence, Italy (2014), pp. 1–6. https://doi.org/10.1109/IEVC.2014.7056197
5. A.A. Pop, M. Radulescu, H. Balan, H. Kanchev, Electromagnetic torque capabilities of axial-flux and radial-flux permanent-magnet machines, 2013 in *4th International Symposium on Electrical and Electronics Engineering (ISEEE)*, Galati, Romania (2013), pp. 1–4. https://doi.org/10.1109/ISEEE.2013.6674365
6. Q.A.S. Syed1, H. Kurtović, I. Hahn, Double stator and single rotor type single-phase flux switching axial flux permanent magnet motor, in *20th International Conference on Electrical Machines and Systems (ICEMS)* (2017), pp. 11–14
7. R. Wilson, R. Gandhi, A. Kumar, R. Roy, Design and analysis of twin-rotor axial flux permanent magnet synchronous motor for electric bicycle using 3D finite element analysis, in *2020 IEEE International Conference on Power Electronics, Smart Grid and Renewable Energy (PESGRE2020)*, Cochin, India (2020), pp. 1–6

Chapter 8
Super-Twisting Algorithm-Based Sliding Mode Control of SMES for Frequency Control in Wind Penetrated Power System

Zahid Afzal Thoker and Shameem Ahmad Lone

Abstract In this paper, frequency control in the wind power penetrated power systems via active power control under varying load conditions is reported. Superconducting magnetic energy storage (SMES) is interfaced with the system to exchange the power mismatch that occurs in generation and demand. To achieve the fast as well as effective exchange of power mismatch and improve the dynamic performance of the system, super-twisting algorithm-based sliding mode controller is proposed on SMES by controlling the firing angle of the converter. Simulation experiments were carried out in MATLAB/Simulink environment, and the results verify the efficacy of the proposed scheme.

Keywords Sliding mode controller · Super-twisting sliding mode controller · Superconducting magnetic energy storage

8.1 Introduction

The increasing energy demands are posing the huge pressure on the conventional energy resources. Renewable energy sources like wind power offer an alternative solution to reduce the impact on the conventional energy-based power systems. However, variations in load and unreliable nature of wind power lead to power quality issues in the power systems. Frequency regulation is the fundamental drawback of the wind penetrated power systems whenever variations in load or wind power occur [1, 2].

Installation of energy storage units with proper power electronic interface could help to achieve the frequency regulation in wind penetrated power systems. Such devices are capable of exchanging the necessary amount of mismatch power that occur between the generation and demand in the system under disturbance conditions [3, 4]. Superconducting magnetic energy storage (SMES) possesses an outstanding ability to exchange and control the power flow in the wind penetrated power systems.

Z. A. Thoker (✉) · S. A. Lone
National Institute of Technology, Srinagar 190006, India
e-mail: thokerzahid_10phd17@nitsri.net

© The Author(s), under exclusive license to Springer Nature Singapore Pte Ltd. 2021
V. Kamaraj et al. (eds.), *Emerging Solutions for e-Mobility and Smart Grids*,
Springer Proceedings in Energy, https://doi.org/10.1007/978-981-16-0719-6_8

Thus, installation of SMES unit could reduce the amount of frequency deviations to a large extent. However, exploring the robust and intelligent control algorithms on SMES, fast and effective exchange of mismatch power could be achieved under disturbance conditions. Frequency control under dynamic load changes using intelligent approach and hybrid energy storage is reported in [5]; however, wind power is taken into account. Conventional control techniques are used in wind power penetrated system to regulate the system frequency, either energy storage units are not installed or else robust control techniques are not included. Gravitational search and genetic algorithms are used in [6] to optimize the PI controller parameters to achieve frequency control in power systems with use of DFIG wind turbines, but such control techniques become practically cumbersome during implementation.

Frequency control via active power control in wind penetrated power systems using energy storage devices like SMES, battery energy storage system, super capacitors is reported [5, 7], but robust and intelligent control schemes are not explored to achieve dynamic performance improvement. Robust control techniques like sliding mode control, adaptive and super-twisting algorithm-based sliding mode control techniques are reported in [8], again either only wind power is used in the system or else energy storage unit alone is incorporated.

In this paper, to improve frequency response, superconducting magnetic energy storage system is interfaced with the wind integrated power system. Super-twisting algorithm along with the conventional sliding mode controller is designed to control the SMES operation. The effective and fast exchange of mismatch power between the system and SMES unit is achieved via power electronic converter by the control of firing angles of the twelve-pulse converter. Finally, comparison of frequency deviations under various disturbances is shown to present the effective performance of the proposed controller-based SMES.

8.2 Modeling and Simulation of the System

The DFIG-based wind penetrated power system model incorporated with SMES unit is shown in Fig. 8.1 [6], where Δf_i, $\Delta P_{\mathrm{G}i}$, ΔP_{TL}, $\Delta P_{\mathrm{L}i}$, $\Delta P_{\mathrm{DFIG}i}$ and $\Delta P_{\mathrm{smes}i}$ represent the deviation in frequency, generator power, tie-line power, load power, DFIG power and SMES power in respective areas.

The deviation in overall active power in each area causes frequency deviation and can be calculated through energy balance relationship given as

$$\Delta f_i(s) = \frac{K_{\mathrm{f}i}}{1 + sT_{\mathrm{f}i}}[\Delta P_{\mathrm{G}i}(s) + \Delta P_{\mathrm{DFIG}i}(s) - \Delta P_{\mathrm{TL}}(s) + \Delta P_{\mathrm{smes}i}(s) - \Delta P_{\mathrm{L}i}(s)] \tag{8.1}$$

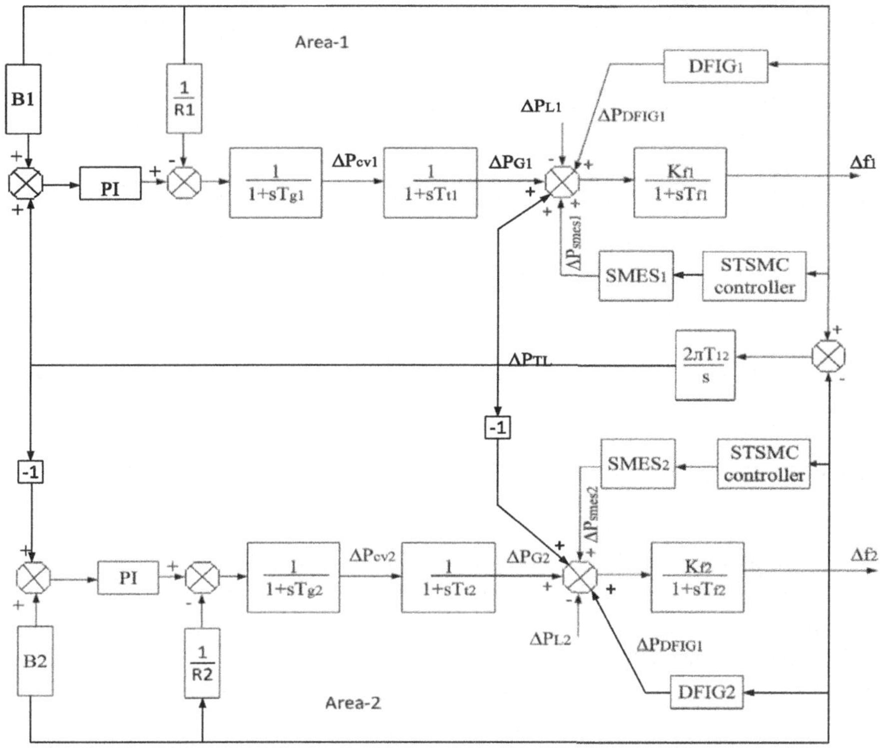

Fig. 8.1 DFIG-based wind penetrated power system model

8.2.1 *Modeling of SMES Unit*

The schematic diagram of superconducting magnetic energy storage system installed in the wind integrated power system is shown in Fig. 8.2, wherein $Y - \Delta/Y$ is a step down transformer which is connected to AC bus, a twelve pulse converter and the super magnetic coil. With the control of converter firing angle α_1 and α_2, the mismatch power exchange between the system and SMES unit is achievable.

In the twelve-pulse converter, the output voltage of the converter in positive, and negative range of values can be obtained by controlling the firing angles α_1 and α_2 of the converter. This feature is achieved by using GTO-based converter, and therefore, following three operation modes are possible:

(i) Charging mode when $\alpha_1 > 0°$ and $\alpha_2 > 0°$, the power flows from system bus to the SMES unit.
(ii) Standby mode when $\alpha_1 = \alpha_2 = 90°$, no power flow is possible via SMES, since the converter output voltage is zero.
(iii) Discharging mode when $\alpha_1 > 90°$ and $\alpha_2 > 90°$, the power flows from the SMES to the system.

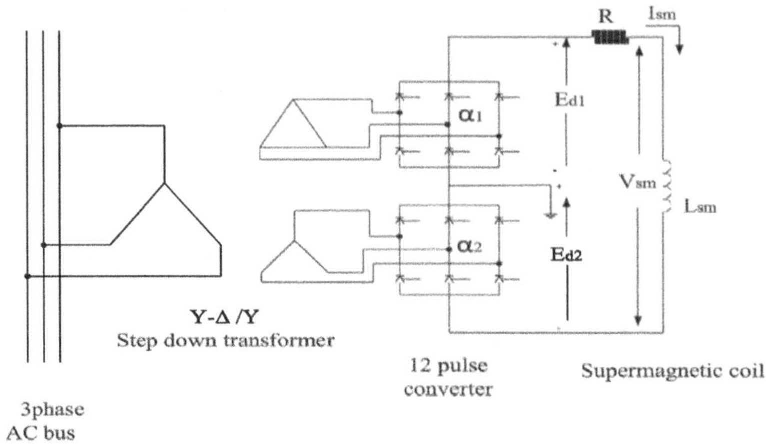

Fig. 8.2 SMES with schematic structure

Depending upon the nature and amount of disturbances, the above operations of the SMES unit can be varied by adjusting the converter firing angles α_1 and α_2 of the converter. To explore the four-quadrant operation of the SMES unit, the converter firing angles can be calculated as [9–11]:

$$\alpha_1 = \cos^{-1}\left(\frac{P_{\text{smes}}}{\sqrt{P_{\text{smes}}^2 + Q_{\text{smes}}^2}}\right) + \cos^{-1}\left(\frac{\sqrt{P_{\text{smes}}^2 + Q_{\text{smes}}^2}}{2V_{\text{sm0}}I_{\text{sm}}}\right) \tag{8.2}$$

$$\alpha_2 = \cos^{-1}\left(\frac{P_{\text{smes}}}{\sqrt{P_{\text{smes}}^2 + Q_{\text{smes}}^2}}\right) - \cos^{-1}\left(\frac{\sqrt{P_{\text{smes}}^2 + Q_{\text{smes}}^2}}{2V_{\text{sm0}}I_{\text{sm}}}\right) \tag{8.3}$$

The converter output voltage on the D–C side can be obtained by the following equation:

$$V_{\text{sm}} = E_{d1} + E_{d2} = V_{\text{sm0}}\cos(\alpha_1 + \alpha_2) \tag{8.4}$$

where E_{d1} and E_{d2} are the *dc* output voltages at the terminal of each converter and V_{sm0} is the no-load maximum value of D–C voltage.

The power both the active and the reactive flowing through the converter is given as

$$P_{\text{smes}} = V_{\text{sm0}}I_{\text{sm}}\cos(\alpha_1 + \alpha_2) \tag{8.5}$$

$$Q_{\text{smes}} = V_{\text{sm0}}I_{\text{sm}}\sin(\alpha_1 + \alpha_2) \tag{8.6}$$

For the SMES coil, the voltage and current relationship is given by

$$L_{sm}\frac{dI_{sm}}{dt} = V_{sm} - RI_{sm} \tag{8.7}$$

where L_{sm} is the inductance of coil, R is the resistance of the coil, I_{sm} is the current following through the coil.

Under continuously varying system frequency, the active power flows through SMES via controlled converter [12] and is given as

$$\Delta P_{sm}(s) = \frac{K_{fc}}{1 + sT_{fc}}\Delta f(s) \tag{8.8}$$

where K_{fc} is the closed loop gain, T_{fc} is time constant of the frequency controller.

8.3 Design of Control Strategy

In order to design the control strategy to achieve the fast and effective active power exchange between the SMES unit and the wind integrated power system by the control of converter firing angle so to reduce the deviations in the frequency response, whenever system disturbances like step/random load changes occur in the system, consider the modeling system diagrams as shown in Figs. 8.1 and 8.3. To design the proposed control scheme, the system dynamics can be expressed as

$$\ddot{x}(t) = f(x(t)) + bu(t) + f_d(t) \tag{8.9}$$

where $x(t) \in R^n$ is the vector of system states, $u(t) \in R^m$ is the vector of control inputs, $f_d(t) \in R^n$ represents the model uncertainty and disturbances in lumped form, $b \in R^n$ is the matrix of input states, $f(x(t))$ is a map of $x(t) \in R^n \rightarrow f(x(t)) \in R^n$. The controller objective is to generate a proper control law in such a way that error in the system state x is tracking the desired/reference error x_{ref}.

Assumption: The uncertainty in the model and the external disturbances represented in lumped form is assumed to be bounded that is $|f_d(t)| \leq \bar{\rho}$, where $\bar{\rho}$ is a positive constant.

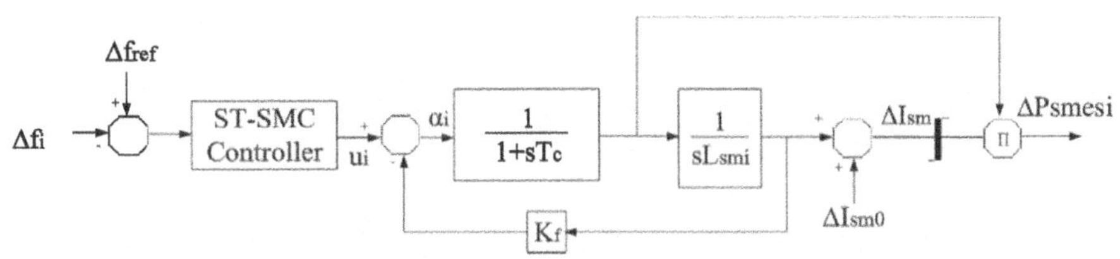

Fig. 8.3 Simulink diagram of SMES with ST-SMC control

8.3.1 Sliding Mode Controller

Sliding mode control offers a simple and robust control scheme. The design procedure involves two steps, the design of a suitable switching manifold and the design of equivalent switching control law that keeps the dynamics of the system onto the designed sliding switching manifold [13, 14]. Let us define a suitable sliding surface as

$$\sigma(t) = \left(\frac{\mathrm{d}}{\mathrm{d}x} + \lambda\right)e(t) = \dot{e}(t) + \lambda e(t) \tag{8.10}$$

where λ is the sliding coefficients and $\lambda > 0$, e and $\dot{e}$ is the state error and derivative of state error, respectively and are defined as

$$e = x_{\text{ref}} - x = [\Delta f_{\text{ref}} - \Delta f] \tag{8.11}$$

$$\dot{e} = \dot{x}_{\text{ref}} - \dot{x} = \left[\Delta \dot{f}_{\text{ref}} - \Delta \dot{f}\right] \tag{8.12}$$

where Δf is the actual deviation in system frequency, Δf_{ref} is the reference deviation in system frequency.

Using (8.9) in (8.10), we have

$$\dot{\sigma}(t) = \ddot{e}(t) + \lambda\dot{e}(t) = (\ddot{x}_{\text{ref}} - f - bu - f_d) + \lambda(\dot{x}_{\text{ref}} - \dot{x}) \tag{8.13}$$

In the traditional sliding mode control (SMC) approach, and the switching law with the constant reaching rate [13] is

$$\dot{\sigma}(t) = -k\text{sgn}(\sigma(t)) \tag{8.14}$$

where $k > 0$ and sgn() is a sign function.

Using (8.14) in (8.13), control law obtained is given as

$$u(t) = \frac{1}{b}(k\text{sgn}(\sigma(t)) + \lambda(\dot{x}_{\text{ref}} - \dot{x}) + \ddot{x}_{\text{ref}} - f - f_d) \tag{8.15}$$

8.3.2 Super-Twisting Sliding Mode Controller

To achieve controller robustness and dynamic performance enhancement of system with SMES, the conventional switching control law is modified by incorporating the super-twisting algorithm [15], the algorithm can be given as

$$\dot{\sigma}(t) = -k_1|\sigma(t)|^{\frac{1}{2}}\mathrm{sgn}(\sigma(t)) - \int k_2\mathrm{sgn}(\sigma(t))\mathrm{d}t \tag{8.16}$$

where $k_1 > 0$, $k_2 > 0$ and sgn() is a sign function.

Using (8.16) in (8.15), super-twisting-based sliding mode control law generated is given as

$$u(t) = \frac{1}{b}\left(-k_1|\sigma(t)|^{\frac{1}{2}}\mathrm{sgn}(t) - \int k_2\mathrm{sgn}(\sigma(t))\mathrm{d}t + \lambda(\dot{x}_{\text{ref}} - \dot{x}) + \ddot{x}_{\text{ref}} - f - f_d\right) \tag{8.17}$$

8.3.3 Stability Analysis

In order to determine the stability analysis of super-twisting-based sliding mode controller while considering the system uncertainty and external disturbances represented by $f_d(t)$ and using ||b|| $\neq 0$, constructing the Lyapunov candidate function as

$$V(t) = \frac{1}{2}\sigma^2(t) \tag{8.18}$$

Taking the derivative of (8.18) and using (8.13), (8.16) and (8.17) along with assumption, we get

$$\begin{aligned}
\dot{V}(t) &= \sigma(t)\dot{\sigma}(t) = \sigma(t)(\ddot{x}_{\text{ref}} - f - bu - f_d + \lambda(\dot{x}_{\text{ref}} - \dot{x})) \\
&= \sigma(t)(-k_1|\sigma(t)|^{\frac{1}{2}}\mathrm{sgn}(\sigma(t)) - \int k_2\mathrm{sgn}(\sigma(t))\mathrm{d}t - f_d \\
&= -k_1\sigma(t)|\sigma(t)|^{\frac{1}{2}}\mathrm{sgn}(\sigma(t)) - \sigma(t)\int k_2\mathrm{sgn}(\sigma(t))\mathrm{d}t - |f_d|(\sigma(t)) \\
&\leq -k_1|\sigma(t)|^{\frac{3}{2}} - |\sigma(t)|\left(\int k_2\mathrm{d}t + \bar{\rho}\right)
\end{aligned}$$

Therefore, with the proper choice of switching gains $k_1 > 0$ and $k_2 \geq \bar{\rho}$, the first derivative of Lyapunov function (8.18) is negative semi-definite with $\dot{V}(t) \leq 0$, and thus the asymptotic stability of the system is guaranteed with $\lim_{t\to\infty}\sigma(t) = 0$, that is in finite time both $\sigma(t)$ and $e(t)$ asymptotically converges to zero.

8.4 Simulation Studies

To evaluate the performance of system with proposed controller-based SMES, the wind integrated power system is subjected to two kinds of load disturbances. Finally, comparison of frequency deviations under each case study is shown with that of conventional (sliding mode controller) SMC-based SMES and PI controller-based SMES.

8.4.1 Case 1: Step Disturbance in Load

Under this case study, the system under consideration is subjected to step disturbance in the system load. At $t = 2$ s, area-1 is set to a load disturbance $\Delta P_L = 0.18$ p.u and also at $t = 2$ s, area-2 is set to a load disturbance $\Delta P_L = 0.2$ p.u. Deviation in frequency in each area occurs, and comparison between the Proportional-integral (PI) controller SMES, sliding mode controller (SMC)-based SMES and super-twisting sliding mode controller (ST-SMC)-based SMES is shown.

From Figs. 8.4 and 8.5, it can been seen that proposed control-based SMES gives superior results in comparison with conventional control-based SMES. Both peak-to-peak deviations in frequency as well as the settling time are reduced with the use of ST-SMC-based SMES in the system. The controller performance with peak-to-peak deviations in frequency and settling time is shown in Tables 8.1 and 8.2.

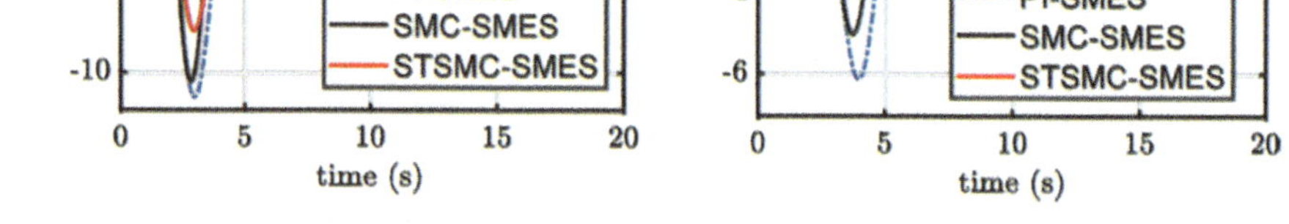

Fig. 8.4 Deviation in frequencies with step disturbance in area-1

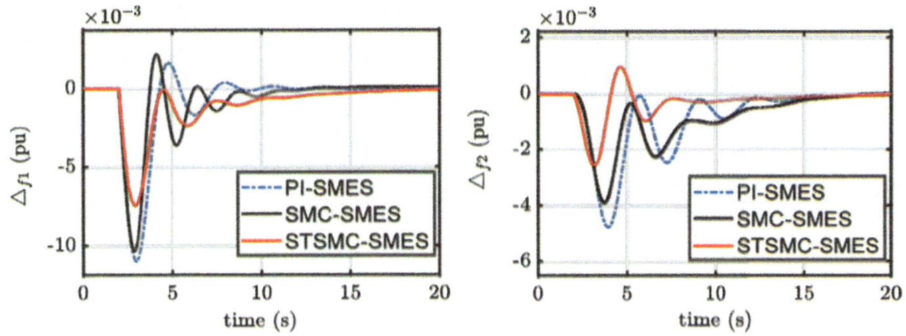

Fig. 8.5 Deviation in frequencies with step disturbance in area-2

Table 8.1 Numerical result of system response with step load change in area-1

Controller	Peak-to-peak Δf (p.u)		Settling time ts (s)	
	$\Delta f1$	$\Delta f2$	$ts1$	$ts2$
PI	0.0213	0.007	15.0	19.5
SMC	0.0211	0.0055	12.5	19.20
ST-SMC	0.0070	0.0042	15.0	14.90

Table 8.2 Numerical result of system response with step load change in area-2

Controller	Peak-to-peak Δf (p.u)		Settling time ts (s)	
	$\Delta f1$	$\Delta f2$	$ts1$	$ts2$
PI	0.0212	0.045	14.75	15.5
SMC	0.0210	0.0375	14.50	15.0
ST-SMC	0.0072	0.0030	15.0	13.5

8.4.2 Case 2: Random Load Variation

To investigate the robustness of proposed controller and achieve the dynamic performance improvement, area-1 of the system under consideration is subjected to a randomly varying load shown in Fig. 8.6.

From Figs. 8.7 and 8.8, it can be observed the under randomly varying load conditions, continuous variation in frequencies occurs. With the use of SMES unit along with proposed control technique, these deviations are reduced sufficiently in comparison with conventional control scheme-based SMES unit.

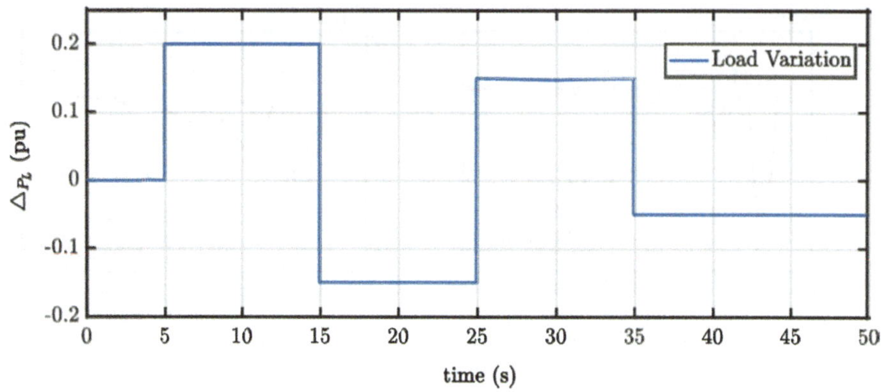

Fig. 8.6 Random load variation

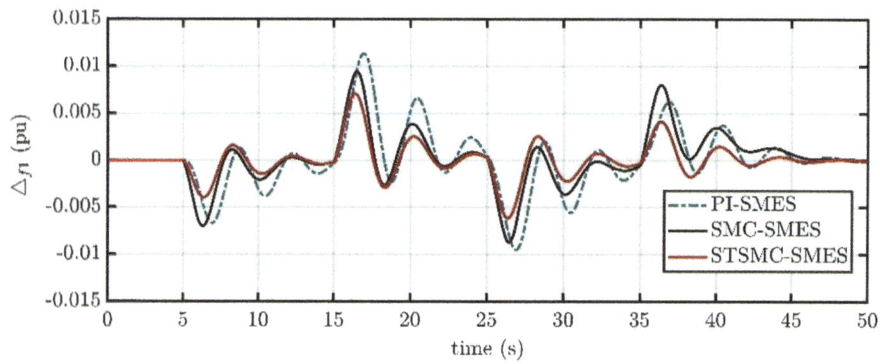

Fig. 8.7 Deviation in frequency of area-1 with disturbance shown in Fig. 8.6

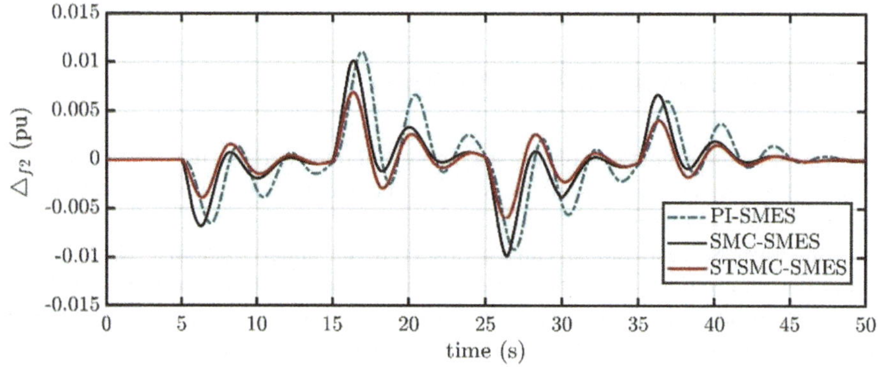

Fig. 8.8 Deviation in frequency of area-2 with disturbance shown in Fig. 8.6

Table 8.3 System parameters

T_{gi}	T_{ti}	R_i	B_i	K_{fi}	T_{fi}
0.2	0.5	0.05	20.6	1	10
0.3	0.6	0.0625	16.92	1	8

Table 8.4 SMES parameters

T_c	I_{sm0}	L_{sm}	K_f
0.5	2	2.65	0.02

Table 8.5 Controller parameters

SMC	STSMC
$\lambda = 1.5$	$\lambda = 1.5$
$k = 8$	$k_1 = 0.3, \quad k = 0.1$

8.5 Conclusions

In this paper, frequency control by active power control in a wind penetrated power system is reported. Superconducting magnetic energy storage system is interfaced with the system to exchange the active power mismatch with the system. With the design of super-twisting algorithm-based sliding mode controller for SMES, fast and effective power exchange is achieved while controlling the firing angles of the twelve-pulse converter. Improvement in frequency response with this controller design under disturbance conditions is achieved in comparison with the sliding mode controller-based SMES and PI controller-based SMES.

Appendix

See Tables 8.3, 8.4 and 8.5.

References

1. Q. Marjela, E. Zeqo, R. Bualoti, A. Mucka, *Impact of the Wind Generation System in Power System Operation* (2010)
2. Z. Chen, *Issues of Connecting Wind Farms into Power Systems* (2005), pp. 1–6
3. S. Associate, S. Swarup, Automatic generation control of two area power system with and without smes: From conventional to modern and intelligent control. Int. J. Eng. Sci. Technol. **3**, 052011 (2011)
4. Y. Mobarak, Robust technique lfc of two-area power system with dynamic performance of combined smes and sssc control. Int. J. Adv. Eng. Technol. **8**, 46–58 (2015)

5. P.V. Ramamurthy, S. Rajan, A. Sivaprakash, An intelligent approach for dynamic load frequency control with hybrid energy storage system. Aust. J. Electr. Electron. Eng. **16**, 1–10 (2019)
6. S.V. Preeti, R. Naresh, Optimal generation control of interconnected power system including fig-based wind turbine. IETE J. Res. **61**, 1–15 (2015)
7. M. Sanduleac, L. Toma, M. Eremia, V.A. Boicea, D. Sidea, A. Mandis, Primary frequency control in a power system with battery energy storage systems, in *2018 IEEE-International Conference on Environment and Electrical Engineering and 2018 IEEE Industrial and Commercial Power Systems Europe (EEEIC/ICPS Europe)* (2018), pp. 1–5
8. K. Liao, Y. Xu, A robust load frequency control scheme for power systems based on second-order sliding mode and extended disturbance observer. IEEE Trans. Industr. Inf. **14**(7), 3076–3086 (2018)
9. M. Elsisi, Control mechanisms of energy storage devices (ch. 3), in *Energy Storage Devices*, ed. by M. T. Demirkan, A. Attia (Rijeka:IntechOpen, 2019)
10. M.G. Rabbani, J.B.X. Devotta, S. Elangovan, Application of simultaneous active and reactive power modulation of smes unit under unequal/spl alpha/-mode for power system stabilization. IEEE Trans. Power Syst. **14**(2), 547–552 (1999)
11. Z.A. Thoker, S. Lone, Voltage and frequency control of wind–diesel power system through adaptive sliding mode control of superconducting magnetic energy storage, in *Wind Engineering* (2020)
12. C. Wu, Y. Lee, Application of simultaneous active and reactive power modulation of superconducting magnetic energy storage unit to damp turbine-generator sub synchronous oscillations. IEEE Trans. Energy Convers. **8**(1), 63–70 (1993)
13. U. Itkis, *Control Systems of Variable Structure: A Halsted Press Book* (Wiley, 1976)
14. M. Khosraviani, M. Jahanshahi, M. Farahan et al., Load–frequency control using multi-objective genetic algorithm and hybrid sliding mode control-based SMES. Int. J. Fuzzy Syst. **20**, 280–294 (2017)
15. Z. Feng, J. Fei, Design and analysis of adaptive super-twisting sliding mode control for a microgyroscope. PLoS ONE **13**(1), (2018)

Chapter 9
Evolution and Recent Advancements in Electric Vehicle (EV) Technology

P. Aruna and V. Vasan Prabhu

Abstract This paper presents the main components of electric vehicles (EV) and the evolution of technologies used in them. The electric vehicles are gaining importance globally as it is seen as a factor of reducing air pollution and smog. This paper focuses on advantages of EVs compared to internal combustion engine (ICE)-based vehicles, different types of EVs and their greenhouse gas emissions, different types of EV motors and their developments, evolution of batteries and their advanced technologies, different types of battery management system (BMS), different types of powertrains and challenges encountered by different types of EVs. EVs are diversified and sophisticated, with different elegance options, but all with the reliability and power of conventional, gasoline-powered vehicles. The main aim of this paper is to give a general image of the present EV innovation and the ways for further improvement tó aid future researches in this area.

Keywords Electric vehicle · Traction motors · Battery management system · Powertrain technologies

9.1 Introduction

Afore the Industrial Revolution, our vitality needs were unassuming. For heat, we relied on the sun and burned wood, straw and dried dung when the sun failed us. For transportation, the potency of the wind in sea and in land, horses took us to every corner of the world [1]. After 1800 A.D, many inventions were made in the transport and subsystems of it, and hence, the usage of electric carriages came into

P. Aruna (✉)
Assistant Professor, Department of Electrical and Electronics Engineering, Anand Institute of Higher Technology (Affiliated to Anna University), Old Mahabalipuram Road, Chennai 603103, India
e-mail: arunap.eee@aiht.ac.in

V. Vasan Prabhu
Assistant Professor, Department of Automobile Engineering, SRM Institute of Science and Technology (Deemed to be University), Kattankulathur, Chennai, Tamil Nadu, India
e-mail: vasanprv@srmist.edu.in

© The Author(s), under exclusive license to Springer Nature Singapore Pte Ltd. 2021
V. Kamaraj et al. (eds.), *Emerging Solutions for e-Mobility and Smart Grids*,
Springer Proceedings in Energy, https://doi.org/10.1007/978-981-16-0719-6_9

existence. In 1900 A.D, the ICE-based vehicles became popular and brought a new era in transport. Due to the low price of petrol and diesel, the usage of two wheelers become larger nowadays which paved the way for air pollution and lot of health problems. So, the automobile industries shifted their attention towards EVs after the invention of lithium ion batteries. Hence, many researchers are focusing on different types of EVs which are classified based on their energy sources. Also, the research papers in electric vehicles are aimed at idealizing and developing the batteries with high density of energy and longer life, fuel cells and its types, battery management system, traction motors, innovative power management and control systems with regenerative braking. This paper presents the literature survey of dark period and come back of EV, comparison between EV and other modes of transport based on energy sources, motor drive technologies and comparison of efficiency and losses of different EV motors, history of batteries, topologies of battery management system, future batteries, powertrain technologies and its advancement and challenges faced by EVs.

9.1.1 Dark Period of Electric Vehicles

As the motor vehicles gained popularity around 1900 A.D, the electric cars were greater in number when compared to diesel/petrol cars. The development of electric starter for petrol cars eliminated the conventional drawback of utilizing a hand crank to get the car moving. After the huge production of Model T by Henry Ford, the era of electric car had come to an end as the price of petrol car is half of the price of an electric car. So, the manufacturing of electric cars was stopped by the year 1935 [2].

9.1.2 Come Back of Electric Vehicles

There was resurgence in electric cars due to the shortage of oil in Gulf countries in 1970s and 80s. With the accessibility and cost of oil being demonstrated to be progressively unstable, people could optically recognize the possible advantages of battery-powered cars. The chance of getting aids and strict regulations made the auto manufacturers of USA, Japan and Europe to develop EVs. Most of the companies in Japan and Europe have developed their EVs around 1960s. The experimental EVs like Electrovair in 1966, Electrovan in 1968, Electrovette in 1979 were launched by General Motors. Following the general motors, many companies started producing different kinds of EVs in late 90s.

9.1.3 Importance of Electric Vehicles

Currently, the global population is 6 billion, and if it increases with the current trend, it may become 10 billion by 2050. This will pave the way for increase in the usage of vehicles 700 million (as per reports in 2000) to 2.5 billion by 2050, and if all these vehicles are IC engine-based vehicles, then most likely all the cities will be covered with permanent smog with extreme air pollution. [3]. This will bring severe health issues. Hence, it is essential to shift the attention towards pollution-free transport. One of the promising solutions is sustainable transport. It means the use of low or zero emission vehicles for public transport. Hence, the captivation of fossil fuels will be less as it is the major source of pollution. So, to better apprehend the benefits and shortcomings of EV, the comparison among different types of EV and ICE-based vehicles can be done on following parameters [4–6].

The parameters considered for comparison are energy sources; pollution, energy diversification, efficiencies, performance capital and operating cost are listed in Table 9.1.

The graph shown in Fig. 9.1 illustrates about the emissions of different types of vehicles. On analysing the figure, it is evident that ICEV-based vehicles have greater percentage of emissions compared to BEV. The other types of vehicles such as HEV, PHEV and REV are in the middle of these two ICEV and BEV.

In India, Bharat Stage regulation permits automobile manufacturers to do research for minimizing the emission of greenhouse gases [7]. The energy diversifications of electric vehicles are given in Fig. 9.2.

The IC engine vehicles normally use liquid fuels and gaseous fuels which are obtained from either oil or natural gas. The pure electric vehicle such as BEV and FCEV uses electricity and hydrogen as energy carters, respectively. Thus, EVs have more benefits compared to IC engine-based vehicles. Among the various types of electric vehicles, PHEV is the most energy diversified as it can run on both liquid fuels and electricity [8].

9.2 Motor Drive Technologies of Electric Vehicles

9.2.1 Evolution of EV Motors

The interesting facts about EV motors are that in 1831, Joseph Henry used electromagnetism for oscillatory motion of beam motor, and this was considered as one of the most primitive motors of the present DC motor. This invention has shown the way for the introduction of the first actual electric motor in 1834 by Thomas Davenport. Following this, Frank J. Spague developed a non-sparking constant-speed motor fixed with brushes. This motor maintained constant speed under varying loads [9].

Table 9.1 List of parameters

Energy sources: The different types of energy sources are liquid fuels, gaseous fuels, battery, ultra-capacitors and ultra-flywheels	
Liquid fuels	Petrol and diesel are called liquid fuels. The specific energy of both petrol and diesel is similar. But the diesel has higher energy content per unit volume as it is denser than petrol. The diesel is more economical compared to petrol due to higher energy density
Gaseous fuels	Compressed natural gas (CNG) and hydrogen are called gaseous fuels. The CNG has lower energy density and higher specific energy compared to liquid fuels. Meanwhile, hydrogen requires storage at a very high compressed state, and therefore, it requires additional accessories around its fuel tank. On the other hand, it has higher energy density compared to petrol, diesel and CNG. Subsequently, there is an exertion to make hydrogen-based fuel cell electric vehicles as a commercial feasibility
Battery	Batteries convert chemical energy directly into electrical energy. Among various types of batteries, let us consider the lithium ion battery. Lithium ion battery has 25 times lower energy density and 75 times lower specific energy compared to petrol. Thus, the mass and volume of lithium ion battery will be 75 times and 25 times more compared to petrol for the same energy content. This is the major drawback of batteries compared to liquid fuels.
Ultra-Capacitors	Ultra-capacitors store the energy in electrostatic form, and it provides very high specific power. But it can store only constrained energy
Ultra-flywheel	Ultra-flywheel stores the energy in mechanical form at high speeds, but it has drawbacks of reliability, and the safety concern has to be considered around it when it is utilized inside the electric vehicle
Pollution: There are different types of pollutants discharged from the vehicles which are the main reasons for air pollution and smog. Some of them are particulate matter (PMx), greenhouse gases, NOx gases, volatile organic compounds (VOC), total hydrocarbons and SOx gases.	

(continued)

Table 9.1 (continued)

Particulate matter	These are the particles released as a part of combustion cycle which are drastically in the range of micrometres. Hence, they cannot be sifted by human throat and nose, and thus, it influences heart, lungs and brain. The particulate matters are released more by diesel engine than petrol engine. PM particles less than 10 μm are known as PM10 are very dangerous and if breathed in, it can cause extreme harm to health
Greenhouse gases (carbon dioxide, methane and nitrous oxide)	Carbon dioxide, Methane and Nitrous Oxide are clubbed and alluded as PM2.5. These gases are known as greenhouse gases. Because of these gases, the infrared radiations are captured in the atmosphere and cause greenhouse effects such as environmental change and a worldwide temperature alteration
NOx gases, VOC and total hydrocarbons	These are responsible for creating ground level ozone layer. If the ozone layer is present at the ground level then humans will be affected by major respiratory diseases such as asthma and lung inflammation. The diesel engine is the key factor of NOx gases. However, these NOx gases can be filtered using urea filters

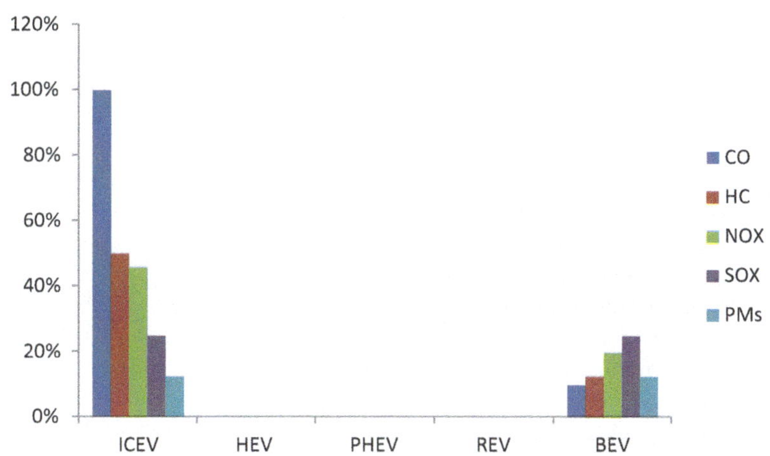

Fig. 9.1 Emissions of different types of vehicles

One of the remarkable inventions in scientific world is the induction motor. Speed regulation, high efficiency and the significant distance dispersion of power conceivable make the induction motor more popular in industries. The induction motor was first invented by the Nikola Tesla in 1887.

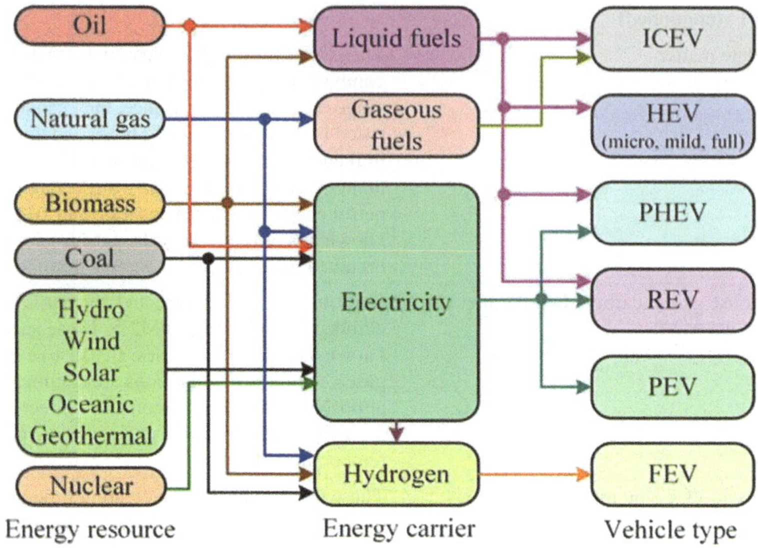

Fig. 9.2 Energy diversification of electric vehicles

In EVs, motor plays a vital role in the drive train. The evolution of EV motors is depicted in Fig. 9.3. The factors given below should be considered for selecting the suitable motor for EVs.

9.2.2 *Main Requirements of an EV Motor*

- Maximum torque and high speed.
- High power and energy density.
- High speed constant power region.
- Low speed constant torque region.
- High reliability and robustness.
- Good voltage regulation in wide speed and
- High efficiency.
- Controllability, steady state and dynamic response.

9.2.3 *Classification of EV Motors*

EV motors are generally classified into brushed DC motors and brushless motors. Brushes and commutator will be present in brushed DC motors and are absent in brushless motors. Generally, all DC motors are commutator-based motors. As the

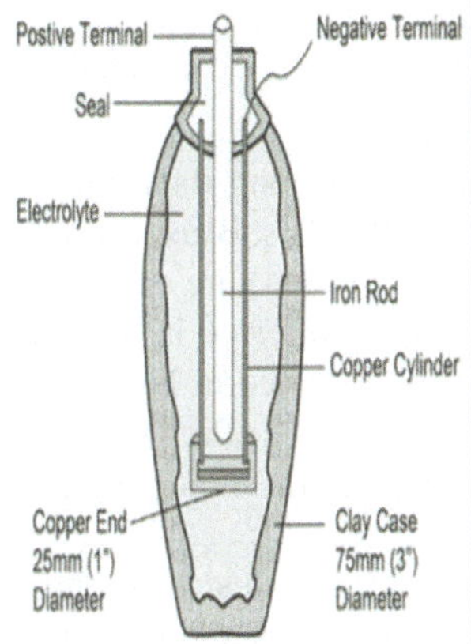

Bhagdad battery in 250 B.C

Battery by Benjamin franklin in 1748

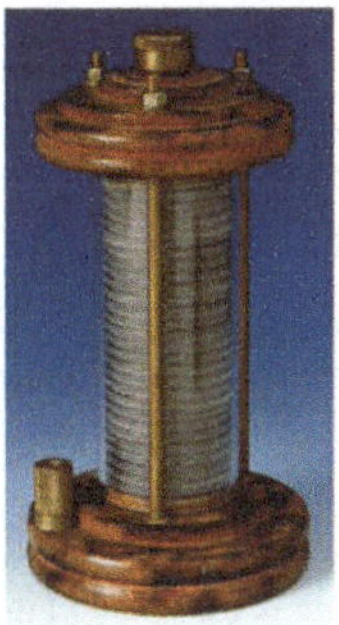

Voltage Pile by
Alessandro Volta in 1799

First rechargeable battery by Plante in 1859

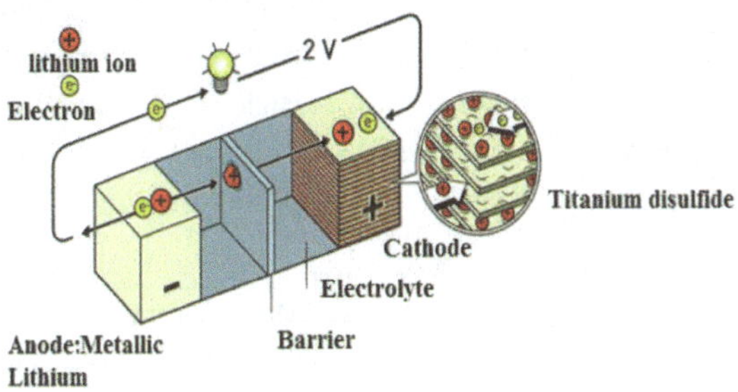

Fig. 9.3 Evolution of EV batteries

construction of DC motors provides symmetrical control of flux and torque, they offer very simple control. Furthermore, the DC motors provide high torque only at low rpm, and the efficiency is only 75–80%. Also, the DC motors require regular maintenance, and life span is short due to the deterioration of brushes. Thus, DC motors are not taken into consideration in EVs [10–13].

AC motors are normally brushless motors. Among them, induction motors and permanent magnet motors are commonly used in EVs. Brushless DC motors (BLDC) and permanent magnet synchronous motors (PMSM) are categorized under permanent magnet motors. BLDC motors are mostly preferred in two wheelers. Thus, the induction motors and PMSM are gaining importance in EVs as they provide high torque at all speed, require less maintenance, and the efficiency is between 85–90%. The comparisons of brushless motors based on different parameters are shown in Table 9.2.

Among these motors, induction motor is mostly desired in industries as it offers maintenance free operation and long life. In early Tesla model cars, the rotor copper bars were used in cage rotor to achieve 85 to 91%. But now; the automobile industries like Tesla, Nissan, Kia motors, BMW have shifted their attention towards permanent magnet motors as these motors are providing high speed-torque characteristics, high

Table 9.2 Comparisons of brushless motors based on different parameters

Items		Brushless motors		
	Induction motor	Permanent magnet (PM) motors		
		Surface mounted PM motor	Interior buried PM motor	PM synchronous reluctance motor
Permanent magnet	No permanent magnet	Magnets are fixed on the exterior of the rotor	Magnets are fixed on the interior of the rotor	Fewer magnets are fixed when compared to IPM
Torque generation	Rotor flux induced by stator flux	Magnetic torque	Magnetic torque and reluctance torque	High reluctance torque than magnetic torque
Efficiency	High	High	High	High
Cost	Low	High	High	High
Speed	Good for high speed	Mechanically weak, particularly at high speed	Good for high speed	Good for high speed
Design	High robust	Less robust	More robust than the SPM	High robust
Field weakening	Field weakening range is high	Not good in field weakening	Field weakening range is medium	Efforts in field weakening necessary

efficiency and power factor. Also, many researches are under progress on the optimization of design parameters and direct torque control of EV motors, and also, the comparisons of efficiencies on different types of EV motors are done in many literatures. Such a comparative analysis has been done on torque (50 Nm) at lower and higher speed in the motors like interior permanent magnet synchronous motor [IPMSM] with 48/8 and 12/8 topologies, induction motor 48 slot/36 rotor bars and switched reluctance motor with 12 stator slot/8 pole [14]. These motors are designed with the same specification of achieving 300 Nm peak torque at 1500 rpm and high torque of 60 Nm at 6000 rpm. Table 9.3 shows the comparison of losses and efficiency of IPMSM, SRM and IM with the torque of 50 Nm at high and low speed [15].

From the above Table 9.3, it is observed that IPMSM with 12 slot/8 pole has the efficiency of 95% at higher speed with lower losses compared to IPMSM with

Table 9.3 Comparison of losses and efficiency of IPMSM, SRM and IM

Type of motor	Torque (Nm)	Speed (rpm)	Permanent magnet loss (W)	Copper loss (W)	Stator copper loss (W)	Rotor copper loss (W)	Core loss (W)	Efficiency (%)
Interior permanent magnet synchronous motor (IPMSM with 48 slot/8 pole)	50	1000	0.01	186	–	–	53	95.5
	50	5000	1.33	767	–	–	552	95
Interior permanent magnet synchronous motor (IPMSM with 12 slot/8 pole)	50	1000	4	88	–	–	66	97
	50	5000	651	418	–	–	299	95
Induction motor (IM with 48 slot/36 rotor bars)	50	1000	–	–	274	215	36	91
	50	5000	–	–	489	458	111	96
Switched reluctance Motor (SRM with 12 stator slot/8 pole)	50	1000	–	283	–	–	379	88.5
	50	5000	–	296	–	–	1346	94

48 slot/8 pole, induction motor and SRM. Hence, IPMSM is providing promising solutions to the EV manufacturers, and hence, it is the most preferred EV motor.

9.3 Batteries and Battery Management System (BMS)

9.3.1 History of Batteries

It is an interesting fact to know about the history of batteries. The history of batteries begins from 250 B.C. approximately. The oldest electric battery was found near Baghdad by a German archaeologist in 1938. This battery comprised of a clay case, an iron rod and a copper cylinder, and some scientists believed that these batteries were used for electroplating gold onto the surface of silver. A set of linked capacitor was described as the battery by the American scientist Benjamin franklin in 1748, and he was the first person to coin the name "battery". The first electric battery "voltage pile" was invented by the Italian scientist Alessandro Volta [16, 17]. Volta set up a fairly untidy pile of zinc and silver plates arranged alternatively and isolated by brackish water splashed cloth. The term volts, unit of voltage is entitled after Alessandro Volta in 1881 in his memory and respect. In 1859, the most generally utilized rechargeable battery was introduced by the French physicist Plante [7]. The evolution of batteries is shown in Fig. 9.3.

The nickel–cadmium (Ni–Cd), nickel metal hydride (NiMH), zinc battery chemistries were developed from 1850 to 1980. The invention of lithium ion battery by John Good Enough in 1980 becomes the eye-opener for the automobile industries to concentrate on electric vehicles. These lithium ion batteries are playing crucial role in all electric vehicles manufactured today as they have high power and energy density, high efficiency, long life span with low self-discharge rate. Lithium ion batteries occupy the superior position over different batteries due to its several advantages [18]. The comparison of lithium ion battery and other batteries is presented in Table 9.4.

The word lithium ion focuses on a group of batteries that have resemblances but have great variation in their chemistries. Such different chemistries of lithium ion are lithium cobalt oxide (LCO), lithium nickel cobalt aluminium oxide (NCA), lithium manganese oxide (LMO), lithium nickel manganese oxide (NMC), and these batteries are comparable as they convey a high volume and are utilized in handy applications. In spite of lower voltage and capacity, lithium iron phosphate and lithium–titanate oxide (LTO) are entirely robust [19]. Table 9.5 describes the features of the most important lithium ion batteries.

NCA is the flawless winner as it has high specific energy than other batteries. In terms of thermal stability and specific power, LMO and LFP are superior. LTO has lower capacity, but it overrides the other batteries in terms of life span and performance in cold temperature performance.

Table 9.4 Comparison of different types of batteries

Battery type	Lead acid	Nickel–cadmium	Nickel metal hydride	Zinc–bromine	Iron–chromium	Lithium ion
Power density Wh/kg	180	150	250–1000	–	70–100	1800
Life cycle	200–300	1500	300–500	>2000	–	500–1000
Energy efficiency (%)	70	60–90	75	80	66	80
Overcharge tolerance	High	Moderate	Low	High	Moderate	Very low
Thermal stability	Less stable	Less stable	Less stable	Less stable	Stable	Highly stable

9.3.2 Future Batteries

In order to create infrastructure based on multi-dimensional energy and to avoid complete dependence on lithium ion batteries, many breakthrough battery technologies are under research to make advancements in powertrain [20]. Table 9.6 shows the different chemistries of lithium ion which may become newer technologies in the near future.

9.3.3 Battery Management System (BMS)

As lithium ion cells are utilized as packs in electric vehicles, which are highly unstable, an observing framework is vital for an individual cell to prevent overcharging and discharging below the threshold value [21]. Thus, BMS gains importance in EVs.

BMS is used in EVs for the following activities:

1. To ensure safe operation and durability of batteries.
2. To disclose state-of-charge and state-of-health.
3. To alert the user during thermal run away of batteries.
4. To alarm the users, when the discharge rate is below the threshold level.

BMS can be classified into centralized BMS, master–slave BMS, modular BMS and distributed BMS [18]. Table 9.7 shows the comparison between the different types of BMS.

Table 9.5 Comparison of different types of lithium ion batteries

Chemistry	Lithium Cobalt oxide	Lithium Manganese oxide (V) ()	Lithium Nickel Manganese Cobalt oxide	Lithium Iron Phosphate	Lithium Nickel-Cobalt Aluminium oxide	Lithium Titanate oxide
Nominal voltage (V)	3.6	3.7	3.6	3.3	3.6	2.4
Voltage at fully charged condition (V)	4.2	4.2	4.2 (or higher)	3.7	4.2	2.9
Voltage at fully discharged condition (V)	3	3	3	2.5	3	1.8
Specific energy (Wh/kg)	150–200	100–150	150–220	90–120	200–260	70–80
Thermal runaway (°C)	150	250	210	270	150	Highly stable at higher temperatures
Applications	Used in mobile phones, laptops and cameras	Used in power tools and medical devices	Used in E-bikes and in medical devices	Used in mobile phones and in watches	Used in industrial and medical applications	Used in UPS and EVs
Comments	High energy, limited power	High power. In order to improve the performance, LMO is mixed with NMC	High capacity and high power	Constant discharge voltage, high power low capacity, very safe	Highest capacity with moderate power	These batteries have fast charging ability, operating in wide range of temperature and longer life. But the cost is very high

Table 9.6 Different chemistries of lithium ion batteries

Chemistry	Lithium-air	Lithium-metal	Solid-state Lithium	Lithium-sulphur Li-S
Type of cathode and anode materials	Cathode-air; Anode-Lithium	Cathode-graphite; Anode-Lithium	Cathode-lithium based oxides Anode-Lithium	Cathode-sulfur; Anode-Lithium
Voltage per cell (V)	1.7–3.2	3.6	3.6	2.1
Comments	Borrowed from "breathing" zinc-air and fuel cell concept	Fast charging capacity and high power	Will be commercialized by 2020	Due to lower cost and higher capacity, these batteries may replace lithium ion batteries in future

Table 9.7 Comparison between the different types of BMS

Centralized BMS	Master-slave BMS	Modular BMS	Distributed BMS
Cells are connected through huge number of wires, and a single controller controls the entire system	Cells are connected through wires and are controlled by slave boards	BMS boards are divided into limited banks and are connected together in a daisy chain. These boards are connected to each cell	Each cell is mounted by a BMS board
	The slave boards are distributed at different sites, and these are controlled by a single master controller	A central controller is connected to BMS boards through a cable	Communication between battery and its controller is done through a cable
The centralized BMS architecture is well suited to battery packs with a small number of cells	1. This architecture divides the tasks of software among the slave boards 2. No additional software requirement, reliabilty are the advantages	Difficult to achieve isolated master–slave communications in electric vehicles	The advantages of this design include its simplicity and high reliability

9.4 Powertrain Technologies

Powertrain embraces of set of components that generate the required power to rotate the wheels and to move the vehicle. In an EV, the powertrain comprises fewer components than internal combustion engine-based vehicles. It includes battery pack, DC-AC converter, electric motor, motor control mechanism and on-board charger

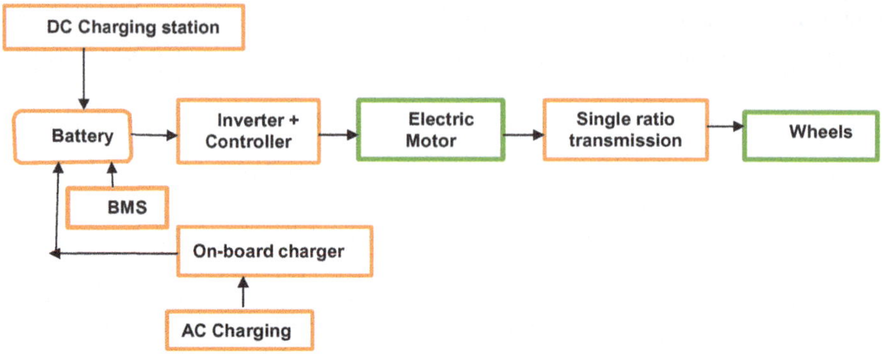

Fig. 9.4 Core parts of powertrain

[14, 22–24]. Along with the core parts, there are also several software and hardware components in the powertrain. Electronic control units (ECUs) are generally software programs combined with the powertrain components to aid data transfer and processing. There are some other several small ECUs like battery management system, DC-DC converter, thermal management system and body control module in an EV that executes particular functions. The communication between different ECUs is normally carried over by CAN protocol. The core parts of powertrain are shown in Fig. 9.4.

A juncture of emanations guidelines, advancements in core technologies and competitions in creating market is making automakers forcefully put resources into new "EV powertrain technologies" and "E-Mobility" solutions. The different architectures of powertrain like front wheel drive, rear wheel drive, all wheel drive and four wheel drive are vying for noticeable quality in different sections of the market [25]. The different types of powertrain technologies are depicted in Table 9.8.

9.4.1 Advanced Powertrain Technologies

It will be always an interest fact to have an inside look of Tesla cars. As Tesla is the pioneer in EVs, the technologies used by them will be always differing from other cars. Recently, two powertrain technologies namely Raven powertrain and Plaid powertrain have been introduced in Tesla cars. The main details about these technologies have been explained below.

9.4.1.1 Raven Powertrain Technology

The Raven powertrain shown in Fig. 9.5 is currently available in the Model S and Model X, but only in performance and long-range guises. This Raven powertrain technology couples the battery available in a Model S or X with the permanent magnet synchronous reluctance motor at the rear side. Compared to induction motor, permanent magnet synchronous reluctance motor is more efficient and powerful than the induction motor. Around 824 lb-ft of torque and 690 horse power is likely to be the total system output [26, 27].

Table 9.8 Different types of powertrain technologies

Item	Picture	Nature of drive
Front wheel drive		• The most common driven cars on the road today • The mass of the engine and transmission is balanced directly on top of the front wheels, providing better traction than rear wheel Drive (RWD) when climbing hills or on slippery roads • Highly stable and efficient [25]
Rear wheel drive (RWD)		• This framework utilizes a long driveshaft that transmits power from the engine to the differential at the back hub • Better weight distribution • More number of components • Better in dry surface [25]

(continued)

Table 9.8 (continued)

Item	Picture	Nature of drive
All wheel drive		• AWD provides better clutch and control under all road conditions • Gives sportier dealing with and footing to a more extensive scope of vehicles • Reduces fuel economy • Increases the complexity and weight of vehicles [25]
Four wheel drive		• Can be used only off-roading and not in everyday life • With 4WD, the vehicle's capacity is part between each of the four wheels, giving the driver the greatest footing and control on the lopsided landscape. This will be helpful when climbing on steep slopes, rolling over rocks and stones or attempting to get the vehicle out of a snowbank [25]

9.4.2 *Plaid Powertrain Technology*

Tesla Company has announced that Plaid mode will be faster than Ludicrous mode which is already available in Tesla models. Three electric motors will be used in Plaid mode will use rather than the two available in other Tesla models [28, 29].

9.5 Challenges Faced by EV

Even though the development in the design, construction and performance of EV is increasing day by day, the challenges faced by EV in the aspect of cost, range, oversizing and energy density cannot be denied [12]. Table 9.9 explains about the challenges faced by different types of EVs.

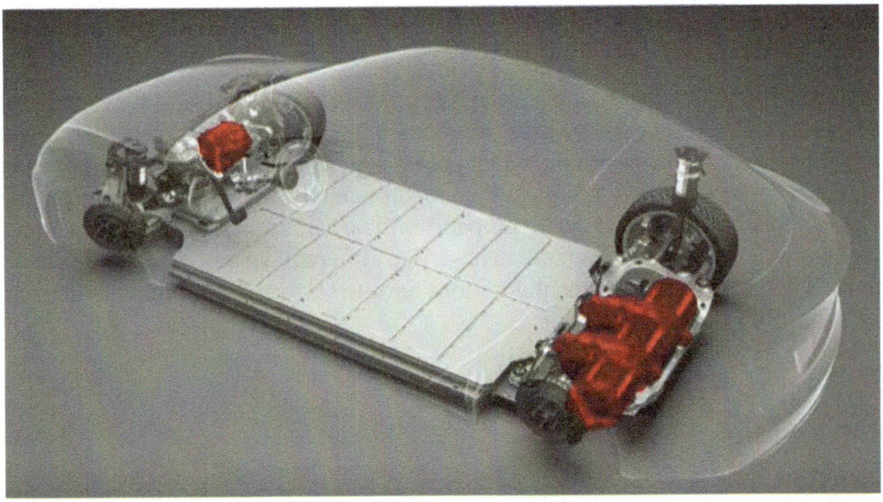

Fig. 9.5 Raven powertrain technology

Table 9.9 Challenges faced by different types of EVs

S. No.	Type of EV	Challenges of the vehicle
1.	Battery electric vehicle (BEV)	(a) Lower energy storage capacity due to low specific energy and low energy density of battery (b) Limited driving range (c) Increase in number of batteries leads to oversizing and high cost (d) Lack of charging infrastructure (e) Charging time can be reduced by increasing the voltage, but it creates stress
2.	Hybrid electric vehicle (HEV)	(a) No zero emissions (b) Dependence on liquid fuels and natural gas (c) Fuel economy can be achieved by variable transmission system, but it requires complex coordinated system, and it creates heavy losses and lot of noise
3.	Grid-able HEV	(a) Capacity of the battery should be high (b) High cost (c) Constraints are similar to HEV
4.	Fuel cell electric vehicle (FCEV)	(a) High cost of fuel cell (b) Lack of infrastructure for hydrogen storage and refuelling (c) Requirement of high safety concerns
5.	Ultra-capacitor electric vehicle	(a) High cost (b) Low specific energy (c) Requirement of battery as hybrid energy source
6.	Ultra-flywheel electric vehicle	(a) High cost (b) Low specific energy (c) Requirement of battery as hybrid energy source (d) Less reliable (e) High safety concern is required as flywheels store mechanical energy at higher speed

9.6 Conclusion

The objective of this paper is to concentrate on the key parts of EV. As EVs become progressively pervasive, pioneering solutions will be essential to create them and their powertrain segments, battery technologies, battery management system, thermal management system and charging system proficiently. The challenges faced by different types of EV presented in this paper will be helpful to concentrate in various domains of EV to improve their efficiency. Also, different types of optimization techniques and power electronic configurations have to be focused in future to simplify the controlling methods.

References

1. Massimo Guarnieri, "Looking Back to Electric Cars", HISTEL conference (2012)
2. H. Rahimi Eichi, U. Ojha, F. Baronti, M.-Y. Chow, Battery management system an overview of its application in the smart grid and electric vehicles. IEEE Ind. Electron. Mag. 932–4529 (2013)
3. G. Du, W. Cao, S. Hu, Z. Lin, T. Yuan, Design and assessment of an electric vehicle powertrain model based on real-world driving and charging cycles. IEEE Trans. Veh. Technol. **68**(2) (2019)
4. K.T. John Wiley & Sons Ltd., Electric Vehicle Machines and Drives—Design, Analysis and Application, Part I
5. I. Husain, *Electric and Hybrid Vehicles Design Fundamentals*, 2nd edn
6. R.V. Krishna kumar, *Electric Vehicles: An Alternate Urban Commuting* (2017)
7. P. Kurzweil, Gaston Planté and his invention of the lead–acid battery-the genesis of the first practical rechargeable battery. J. Power Sour. 4424–4434 (2010)
8. E.A. Grunditz, T. Thiringer, Performance analysis of current BEVs based on a comprehensive review of specifications. IEEE Trans. Transp. Electrification, 270–289 (2016)
9. https://interestingengineering.com/a-brief-history-and-evolution-of-electric-cars
10. P. Naresh Bhatta, H. Meharb, M. Sahajwanib, Electrical Motors for Electric Vehicle—A Comparative Study International Conference on "Recent Advances in Interdisciplinary Trends in Engineering and Applications, SSRN-ELSEVIER (2018)
11. S. Jamshed Rind1, Y. Ren, Y. Hu, J. Wang, L. Jiang, Configurations and control of traction motors for electric vehicles: a review. Chin. J. Electr. Eng. **3**(3) (2017)
12. F. Un-Noor, S. Padmanaban, L. Mihet-Popa, M. Nurunnabi Mollah, E. Hossain, A comprehensive study of key electric vehicle (ev) components, technologies, challenges, impacts, and future direction of development. Energies **10**, 1217 (2017)
13. T. Porselvi, M.K. Srihariharan, J. Ashok, S. Ajith Kumar, Selection of power rating of an electric motor for electric vehicles. Int. J. Eng. Sci. Comput. **7**(4) (2017)
14. Z. Yang, F. Shang, I.P. Brown, M. Krishnamurthy, Comparative study of interior permanent magnet, induction, and switched reluctance motor drives for EV and HEV applications. IEEE Trans. Transp. Electrification **1**, 245–254 (2015)
15. B. Bilgin, A. Emadi, Electric motors in electrified transportation: a step toward achieving a sustainable and highly efficient transportation system. IEEE Power Electron. Mag. **1**(2), 10–17 (2014)
16. https://www.financialexpress.com/auto/car-news/
17. https://www.beatson.co.uk/history-electric-motors/
18. http://liionbms.com/php/wp_spaghetti.php
19. https://batteryuniversity.com/learn/article/types_of_lithium_ion

20. J.B. Goodenough, K.-S. Park, The Li-Ion rechargeable battery: a perspective. J. Am. Chem. Soc. **135**, 1167–1176 (2013)
21. J.B. Goodenough, H. Gao, A perspective on the Li-ion battery. Sci. China Chem. **62**, 1555–1556 (2019)
22. G. Wu et al., Powertrain architectures of electrified vehicles: review, classification and comparison. J. Franklin Inst. Eng. Appl. Math. **352**, 425–448 (2015)
23. I. Boldea, L.N. Tutelea, L. Parsa, D. Dorrell, Automotive electric propulsion systems with reduced or no permanent magnets: an overview. IEEE Trans. Ind. Electron. **61**(10), 5696–5711 (2014)
24. V. Bakshi, V.S. Jape, Drive selection and performance evaluation of electric and hybrid electric vehicles. Int. J. Eng. Res. Technol. (IJERT) **3**(10) (2014). ISSN: 2278-0181
25. https://www.torque.com
26. M. Ehsani, Y. Gao, S.E. Gay, A. Emadi, *Modern Electric, Hybrid Electric, and Fuel Cell Vehicles*
27. https://www.tesla.com/sv_SE/blog/
28. https://arstechnica.com/cars/2019/04/motor-technology-from-model-3-helps-tesla-boost-model-s-range-10/
29. https://www.tesla.com

Chapter 10
Design and Implementation of Boost-Buck DC–DC Converter for Battery Charging Application

Seyezhai Ramalingam, S. Harika, A. Sowmya, N. Ramakrishnan, and S. Purushothaman

Abstract For an electric vehicle, battery plays a major role and needs frequent charging. For this, an efficient DC–DC converter is essential to provide ripple-free and steady output power so that the performance of the battery will not be deteriorated. This paper investigates a myriad of topologies of DC–DC converter and proposes a boost-buck converter for battery charging applications. The performance of the suggested boost-buck converter has been examined and compared with the existing non-isolated topologies. The criterion answered for comparative studies is output voltage and current ripple, input current ripple, and efficiency. From the analysis, it has been implied that the boost-buck configuration has reduced ripple and improved efficiency. Therefore, it is recommended for the battery charger. The simulation studies are executed in MATLAB software. To authenticate the simulation results, a laboratory prototype has been constructed.

Keywords Electric vehicle · Ripple · Boost-buck converter

10.1 Introduction

Carbon emission from the burning of fossil fuel causes air pollution, which leads to global warming that becomes a major threat. Thus, switching over to electric vehicles from an internal combustion engine vehicle is an effective solution for global warming. The electric vehicle has numerous advantages such as zero emission, fewer maintenance requirements, and eco-friendly. Battery, hybrid, and plug-in hybrid electric vehicles are the types of the electric vehicle. Battery-operated electric vehicle is mostly preferred due to its energy efficiency [1–4]. As the non-renewable energy sources are limited in nature and its consumption is faster than its replenishing time, renewable energy sources are widely used for charging the battery. Compared to all

S. Ramalingam · S. Harika · A. Sowmya (✉) · N. Ramakrishnan · S. Purushothaman
Renewable Energy Conversion Laboratory, Department of EEE, SSN College of Engineering, Kalavakkam, Chennai 603110, India

S. Ramalingam
e-mail: seyezhair@ssn.edu.in

© The Author(s), under exclusive license to Springer Nature Singapore Pte Ltd. 2021

V. Kamaraj et al. (eds.), *Emerging Solutions for e-Mobility and Smart Grids*,
Springer Proceedings in Energy, https://doi.org/10.1007/978-981-16-0719-6_10

types of renewable energy sources, solar power has been used in an electric vehicle due to its abundant availability and free from emissions [5, 6]. In solar power system, the solar output power does not meet the battery specification as the single solar cell generates low power and also, the output power varies due to partial shading, dust, and weather conditions. Thus, a DC–DC converter is essential to regulate the solar output power to improve the battery performance. Such converters which act as an interlink between the solar panel and the battery are used. The converter topology employed for battery charging should possess minimum ripple, deliver continuous output power, and be highly efficient. There are different topologies of DC–DC power converters available. The most recurrently used topologies are boost [7], buck [8], buck-boost [9], Cuk converters.

But, the traditional topologies such as boost, buck, and buck-boost have demerits of high output voltage ripple and slow transient response. Many other topologies have been reviewed such as interleaved boost [10], cascaded boost [11], quadratic boost [12], and fly-back converter. In interleaved boost converter as well as in cascaded boost converter, the number of passive component increases, and hence, the size of the converter is bulky. And in the quadratic boost converter, the voltage and current rating of a switch are higher as it integrates the boost converter. Similarly, the fly-back converter also faces the drawback of higher output voltage ripple and higher power losses. Thus, a topology comprising a boost and buck topology has been proposed in this paper.

The proposed boost-buck converter overcomes the problem associated with conventional topologies. The performance of the boost-buck converter is computed and compared with boost and buck topology. The fulfillment parameters are ripple in the output voltage (ΔV_O), ripple in the output current (ΔI_O), and ripple in the input current (ΔI_i). From the studies, it has been inferred that the suggested topology has better performance than the existing topologies. Because the converter employed for battery charging should possess a low ripple in both output voltage and output current, so that, the battery performance and lifetime will be enhanced. The block diagram of the battery charging system is shown in Fig. 10.1. The proposed boost-buck converter steps up the 35 V input voltage to an output voltage 40 V. Two 80 W solar panels are cascaded to obtain 35 V. The output of the converter charges the Li-ion battery rating of 36 V, 60 Ah.

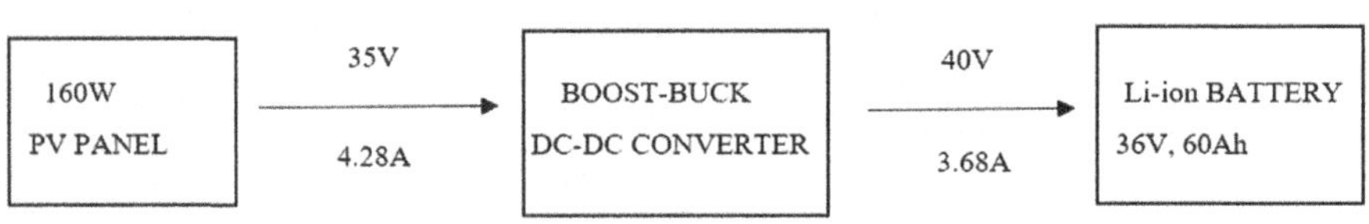

Fig. 10.1 Block diagram of the battery charging system

10.2 PV Modeling

As the single solar PV panel output is low, the solar panels are connected in either series or parallel to meet the load requirement. The mathematical modeling used for the PV model is discussed in this section. Figure 10.2 presents the equivalent circuit of the PV cell.

The output current equation [13] is expressed as

$$\mathrm{I} = \mathrm{I_{ph}} - \mathrm{I_0}[\exp(\mathrm{q}.(\mathrm{V} + \mathrm{I.R_s})/(\mathrm{n.N_s.T.K})) - 1)] - ((\mathrm{V} + \mathrm{I.R_s})/(\mathrm{R_{sh}})) \tag{10.1}$$

where I is the output current, I_{ph} is the photocurrent, I_0 is the reverse saturation current of diode, q is the charge of an electron, V is the cell terminal voltage, R_s is the series resistance, R_{sh} is the shunt resistance, n is the diode ideality factor, N_s is the sum of series connected cell, T is the temperature of the p–n junction, and K is the Boltzmann constant. The insolation and the cell temperature are the major factors that affect the photocurrent [13].

$$I_{\text{ph}} = [I_{\text{sc}} + k_i(T - 298)] \cdot \frac{G}{1000} \tag{10.2}$$

where I_{sc} is the short-circuit current, k_i is the temperature coefficient, and G is the illumination of the PV panel (W/m^2). The diode saturation current varies as a cubic function [13] of the temperature and is given by the following equation

$$I_0 = I_{\text{rs}} \cdot \left(\frac{T}{T_n}\right)^3 \cdot \exp\left[\frac{q \cdot E_{g0} \cdot \left(\frac{1}{T_n} - \frac{1}{T}\right)}{n \cdot K}\right] \tag{10.3}$$

And the reverse saturation current of PV (I_{rs}) is expressed as [13]

$$I_{\text{rs}} = \frac{I_{SC}}{e^{\left[\frac{q \cdot V_{OC}}{n \cdot N_S \cdot T \cdot K}\right]} - 1} \tag{10.4}$$

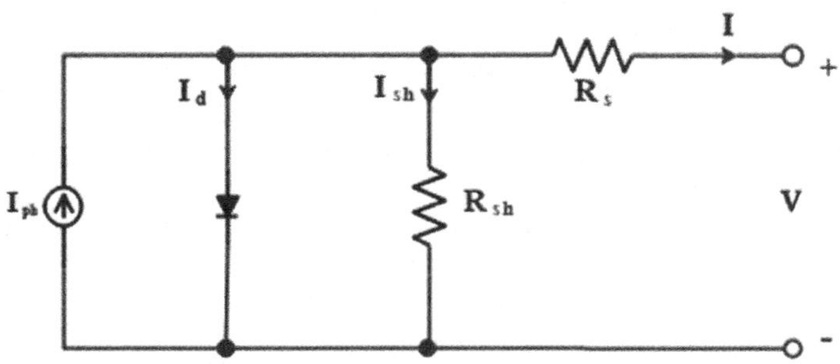

Fig. 10.2 Equivalent circuit of PV cell

where E_{g0} is the energy bandgap of the semiconductor and T_n is the reference temperature of the cell. The I–V and PV characteristics of the 160 W PV module are shown in Figs. 10.3 and 10.4. The PV characteristics for variable irradiation and variable temperature are depicted in Figs. 10.5, 10.6, 10.7 and 10.8.

As shown in Figs. 10.3 and 10.4, the maximum power is achieved at a voltage corresponding to the knee point, which is called as maximum power point (MPP).

From Figs. 10.5 and 10.6, it has been observed that, with increases in insolation level, the current increases which in turn attains the maximum power at 1000 W/m^2.

From Figs. 10.7 and 10.8, it is observed that power decreases with an increase in temperature. The increase in diode current results in decrease in powers.

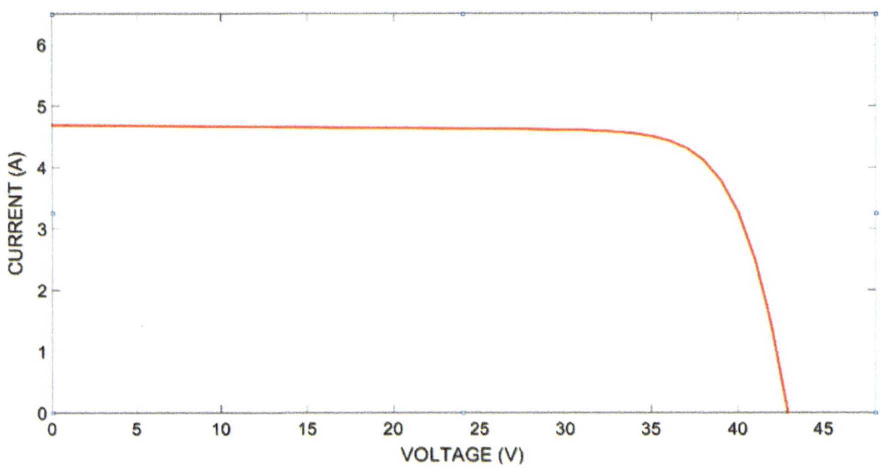

Fig. 10.3 I–V characteristics for 160 W solar PV module

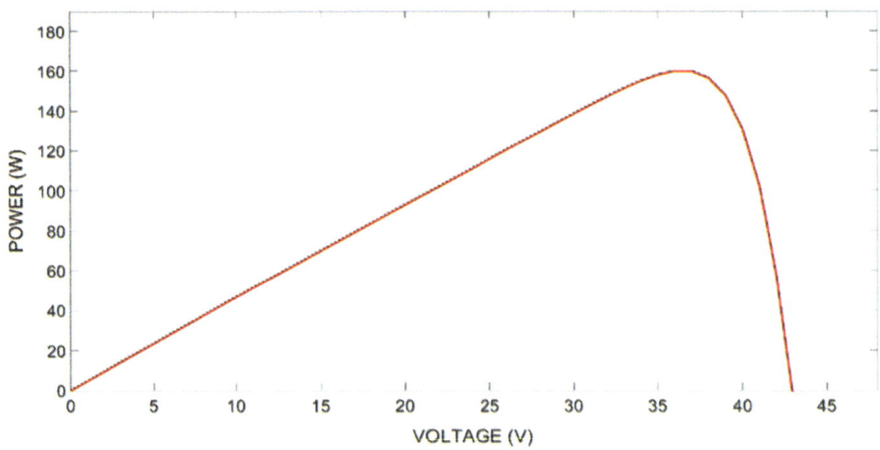

Fig. 10.4 PV characteristics for 160 W solar PV module

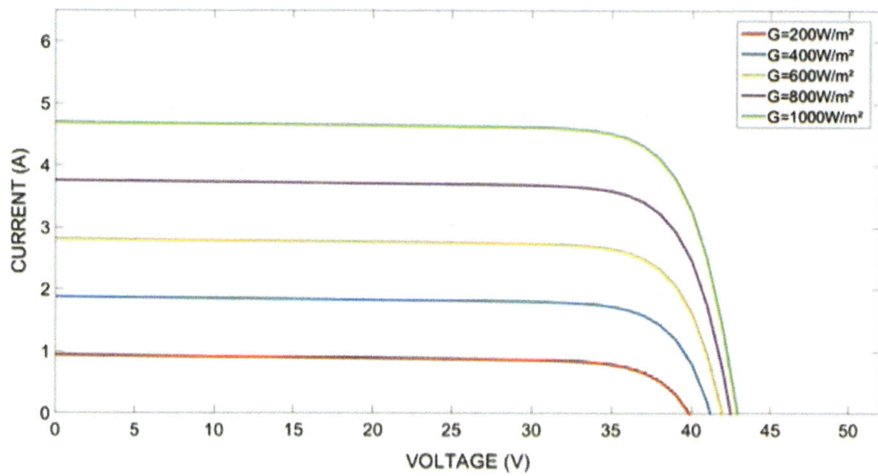

Fig. 10.5 *I–V* characteristics for variable irradiance (*G*)

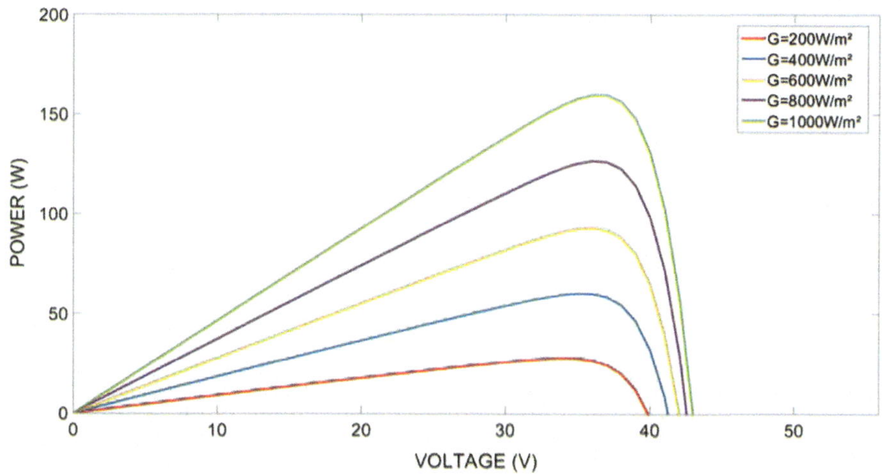

Fig. 10.6 PV characteristics for variable irradiance levels

10.3 Examinations of Topologies of DC–DC Converter

10.3.1 Boost Converter

A boost converter is the most widely used DC–DC topology which performs the step-up operation. By the law of conservation of energy, ideally, the input power is equal to the output power (neglecting the losses). So, the voltage and current change

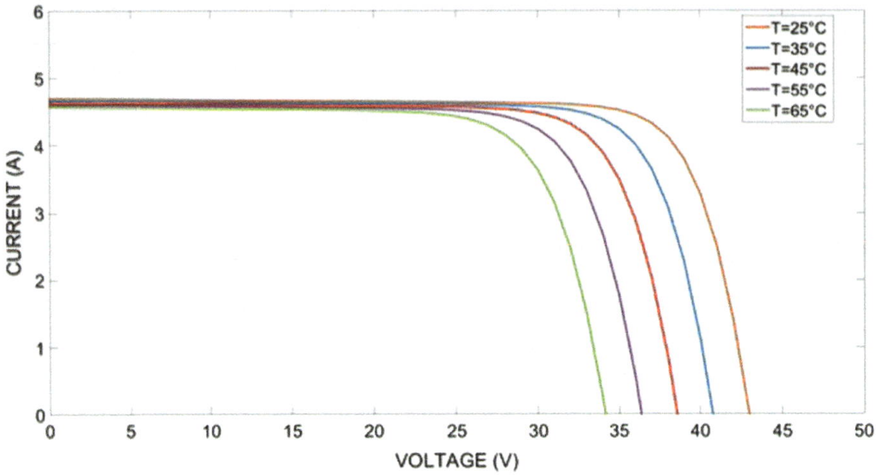

Fig. 10.7 *I–V* characteristics for variable solar temperature (*T*)

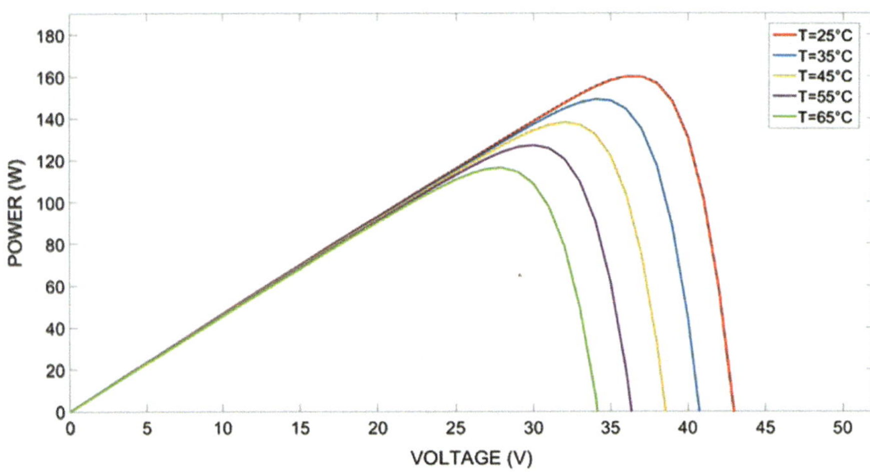

Fig. 10.8 PV characteristics for variable solar temperature

proportionately. It consists of a power electronic switch, diode, and passive elements. The circuit configuration of the boost converter is shown in Fig. 10.9.

When the switch *S* is turned on, the magnetic field builds around the inductor *L*. Due to this, the inductor current increases continuously. When the switch *S* is turned off, the polarity of inductor magnetic field changes; as a result of this, the diode gets for-ward biased and turns on. Now the inductor current decreases linearly. The stored energy in the inductor charges the capacitor as well as powers up the load.

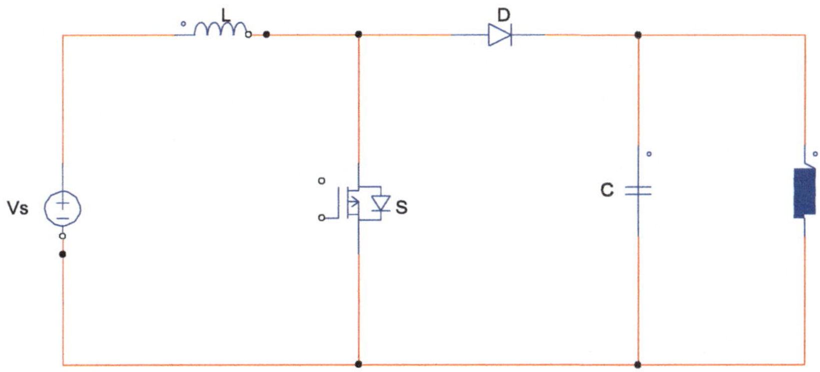

Fig. 10.9 Circuit diagram of boost converter

10.3.2 Buck Converter

In buck converter, the output voltage is lesser when compared to input voltage; therefore, it is a step-down converter. The output current is stepped up simultaneously. Buck converters are widely used in computer applications. Figure 10.10 shows the circuit diagram of the buck converter.

When the switch *S* is turned on, the diode gets reverse biased and the input appears across the load. The current flows through inductor *L*, capacitor *C*, and the load, while the capacitor has been charged during this turn-on period. When the switch *S* is turned off, the polarity of the inductor changes. Now, the diode *D* freewheels the inductor current which flows across the load.

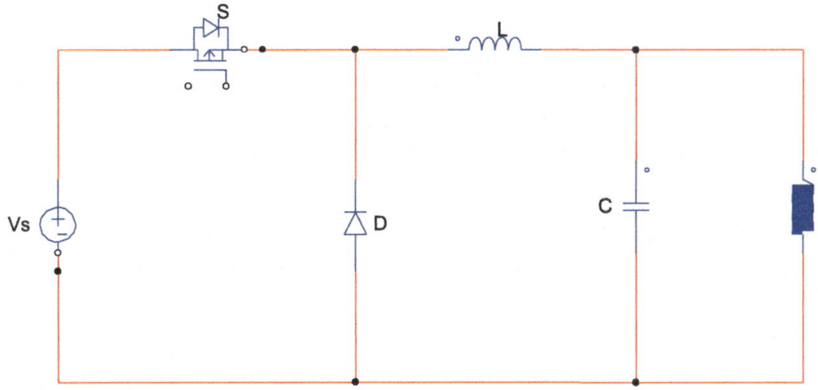

Fig. 10.10 Circuit diagram of buck converter

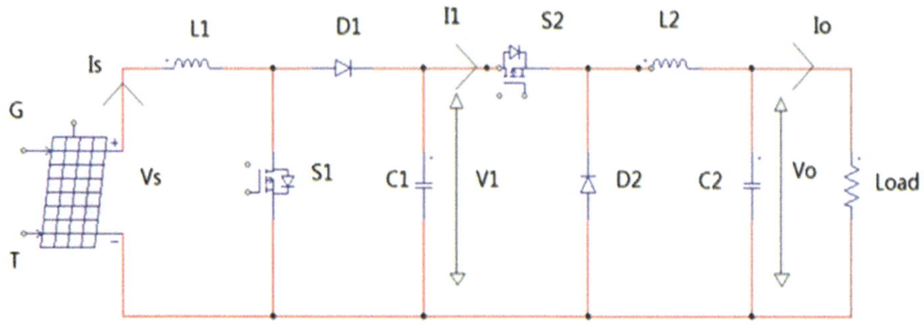

Fig. 10.11 Circuit diagram of boost-buck converter

10.3.3 Boost-Buck Converter

The boost-buck converter can be obtained by integrating the boost and buck converter. Figure 10.11 shows the circuit configuration of the boost-buck converter. With this converter, the output voltage and current ripple are low in comparison with traditional boost and buck converters. In this section, the design equations and working of the proposed converter have been discussed. The operation is of two modes, namely boost mode (mode 1 and mode 2) and buck mode (mode 3 and mode 4).

Mode 1 ($0 < t \le t_{on}$): As shown in Fig. 10.12a, when the switch S_1 is turned on, voltage at the input appears across the inductor L_1 and the magnetic field builds up in the inductor L_1.

Mode 2 ($t_{on} < t \le t_{off}$): When the switch S_1 is turned off as shown in Fig. 10.12b, the inductor L_1 polarity is changed and, thus, forward biases the diode D_1. The energy stored in the inductor charges the load and capacitor, and hence, boosted voltage has been obtained.

Mode 3 ($0 < t \le t_{on}$): When the switch S_2 is turned on as shown in Fig. 10.13a, the current flows through the inductor L_2, capacitor C_2, and the load. The diode D_2 is reverse biased. The capacitor C_2 gets charged, and the capacitor voltage appears across the load.

Mode 4 ($t_{on} < t \le t_{off}$): When the switch S_2 is turned off and diode D_2 is in on condition as shown in Fig. 10.13b, the inductor current decreases linearly and the voltage across the capacitor C_2 appears across the load (Fig. 10.14).

The output voltage for boost converter is expressed as [14].

$$V_1 = \frac{1}{1 - d_1} V_S \tag{10.5}$$

where V_1 is the output voltage of boost converter, V_S is the input voltage of boost converter, and d_1 is the duty cycle of the boost converter. The inductor value is designed using [14]

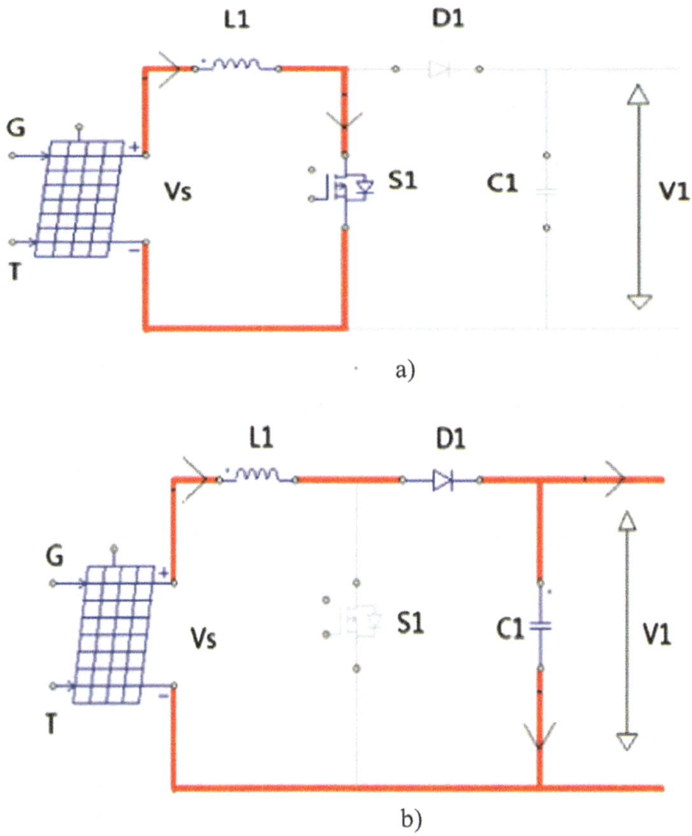

Fig. 10.12 Circuit configuration of boost-buck converter in boost mode: **a** on-state and **b** off-state

$$L_1 = \frac{V_S(V_1 - V_S)}{\Delta I_1 f V_1} \tag{10.6}$$

where is the ripple current of and f is the switching frequency. The design value for capacitor value is [14]

$$C_1 = \frac{I_1 d_1}{f \Delta V_1} \tag{10.7}$$

where ΔV_1 is the ripple voltage of V_1 and I_1 is the output current of boost converter.

The output voltage equation of the buck converter [15] can be expressed as

$$V_0 = V_1 * d_2 \tag{10.8}$$

where V_0 is the output voltage of buck converter and d_2 is the duty cycle of the buck converter. The value of inductance is given by [15]

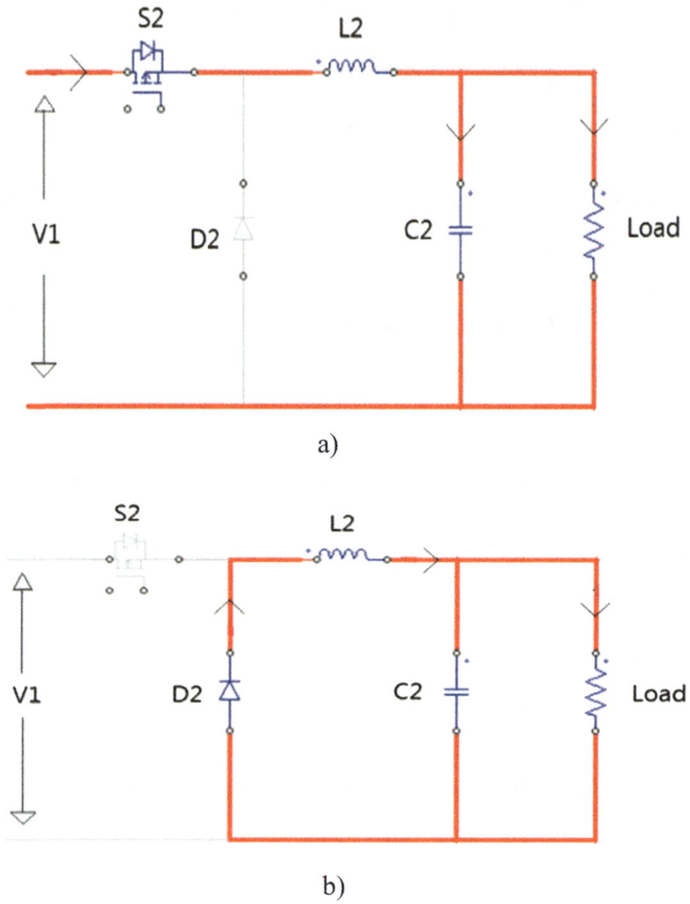

Fig. 10.13 Circuit configuration of boost-buck converter in buck mode: **a** on-state and **b** off-state

$$L_2 = \frac{d_2(V_1 - V_0)}{f \Delta I_2} \tag{10.9}$$

The value of capacitance is [15]

$$C_2 = \frac{V_1 d_1 (1 - d_2)}{8 L_2 \Delta V_2 f^2} \tag{10.10}$$

The load R is given by

$$R = \frac{V_0}{I_0} \tag{10.11}$$

where I_0 is the output current of the buck converter.

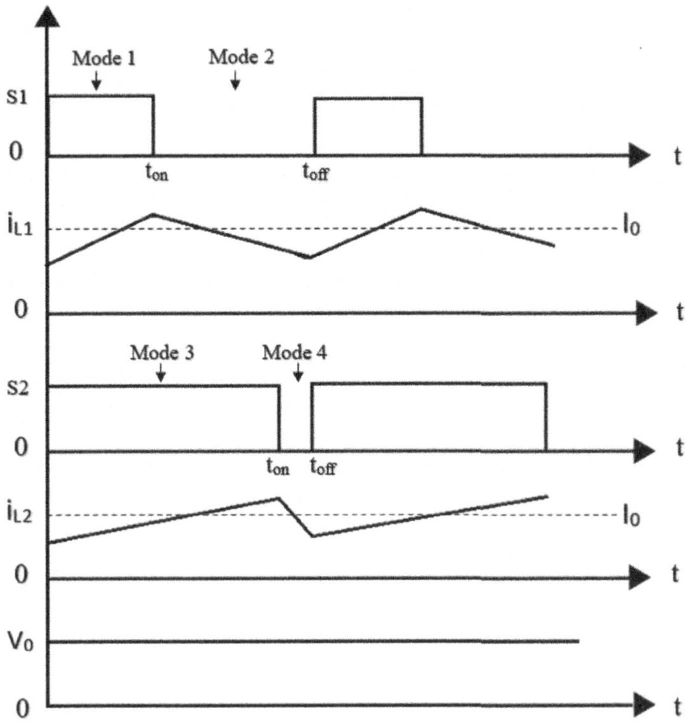

Fig. 10.14 Waveform of boost-buck converter

10.4 Simulation Results

The simulation studies of the battery charger circuit have been performed in MATLAB/Simulink. The chosen boost-buck converter has been interfaced with 160 W PV panel. Two 80 W panels with $V_{oc} = 21.5$ V, $I_{sc} = 4.70$ A, $V_{mp} = 17.5$ V and $I_{mp} = 4.28$ A are cascaded. The total rating of the PV module is $P_{mp} = 160$ W, $V_{oc} = 43.0$, $I_{sc} = 4.70$ A, $V_{mp} = 35$ V and $I_{mp} = 4.28$ A. The specification of the charger circuit is listed in Table 10.1.

The input and output voltage waveform, and input and output current waveform of the boost-buck converter are shown in Figs. 10.15 and 10.16.

From Fig. 10.15, it can be inferred that the input voltage (V_S) of 35 V is boosted to 41 V (V_1) for the duty ratio of 0.16 and stepped down to 40 V (V_0) for the duty ratio of 0.95. The output voltage ripple is 0.051 V.

The input current (I_s) of 4.5 A is stepped down to 3.5 A (I_1) for the duty ratio of 0.16 and boosted to 3.7 A (I_0) for the duty ratio of 0.95 as shown in Fig. 10.16. The output current ripple is 0.055 A.

The inductor voltage waveforms and switch voltage waveforms of the boost-buck converter are depicted in Figs. 10.17 and 10.18. From the results, it has been

Table 10.1 Circuit parameters of boost-buck converter

Parameters	Values
Maximum power (P_{max})	160 W
Optimum operating voltage (V_{mp})(V_s)	35 V
Optimum operating current (I_{mp})(I_{in})	4.28 A
Open-circuit voltage (V_{oc})	21.5 V
Short-circuit current (I_{sc})	4.70 A
Output voltage (V_0)	40 V
Output current (I_0)	3.68 A
Switching frequency (f)	10 kHz
Duty cycle of boost converter (d_1)	0.16
Duty cycle of buck converter (d_2)	0.95
Inductor (L_1)	541.62 μH
Capacitor (C_1)	27.35 μF
Inductor (L_2)	172 μH
Capacitor (C_2)	7.25 μF
Resistor (R)	10.86 Ω

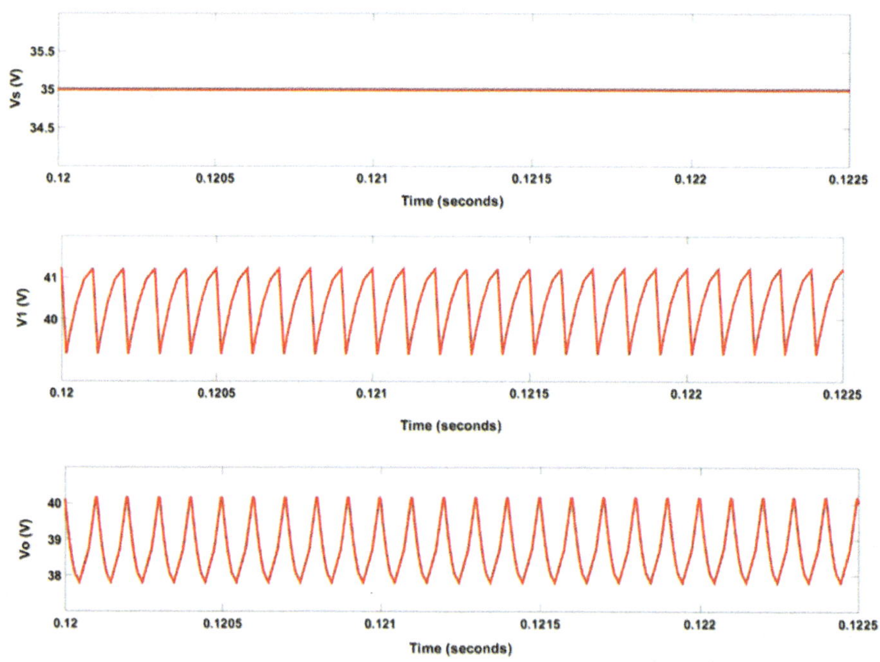

Fig. 10.15 Voltage waveforms of boost-buck converter

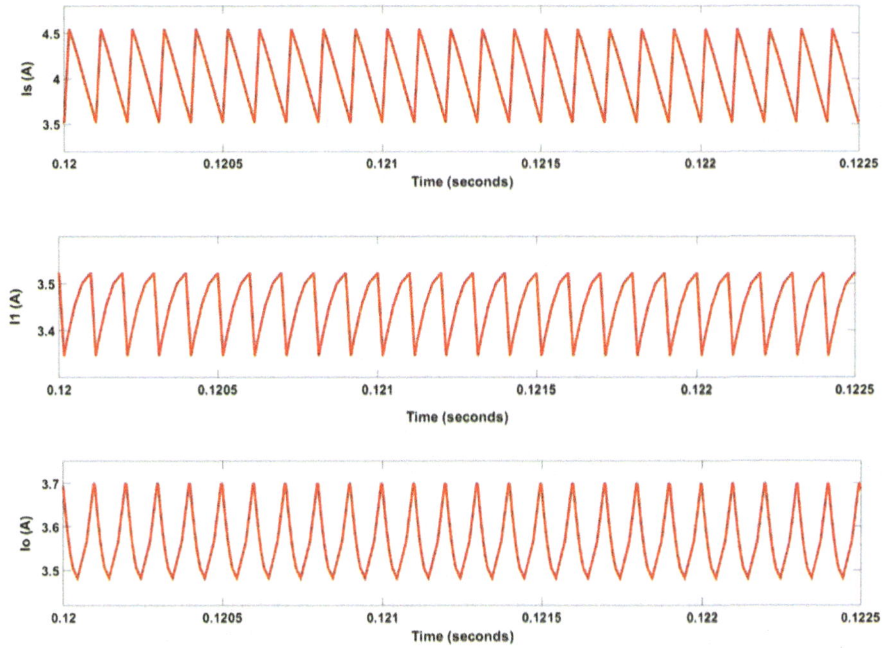

Fig. 10.16 Current waveform of boost-buck converter

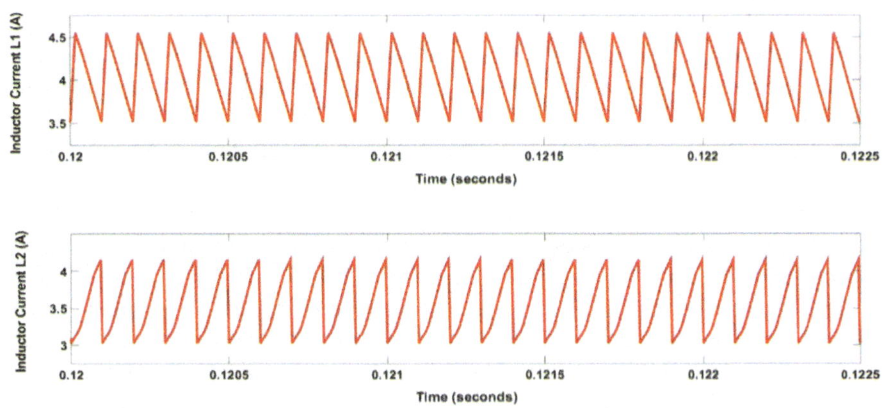

Fig. 10.17 Inductor waveform of boost-buck converter

implied that the proposed converter has reduced output voltage ripple and output current ripple. And, the efficiency of the suggested boost-buck converter is 93.8% which is higher than the boost (90.07%) and buck (90.8%) converters. Hence, it is recommended for a battery charging application.

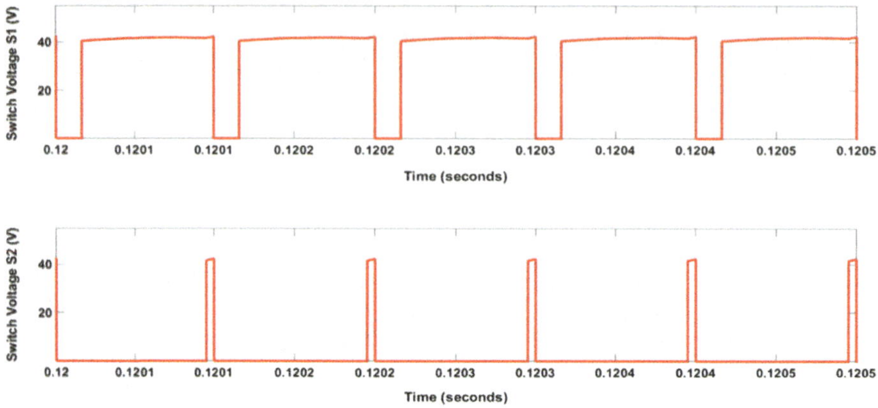

Fig. 10.18 Switch voltage waveform of boost-buck converter

10.5 Analysis of DC–DC Converter

The performance of the suggested converter topology has been evaluated for different duty ratios and compared with existing boost and buck converters. The parameters taken into account for analysis are output voltage ripple (ΔV_0), output current ripple (ΔI_0), and input current ripple (ΔI_i). The specification of the converter has been selected based on the equations described in Sect. 3.3. The designed values are the same for all the three converters.

From Fig. 10.19, it has been inferred that the proposed topology has low output voltage ripple of 0.0392 at 40% duty ratio against the boost (0.127 at 40% duty ratio) and buck converter (0.634 at 40% duty ratio).

From Fig. 10.20, it can be observed that the input current ripple of the boost-buck converter is low when compared to the traditional boost and buck converter. The boost-buck converter has input current ripple of 0.2 at 40% duty ratio, and boost and buck converters have input current ripple of 0.42 and 1.96 at 40% duty ratio, respectively.

From Fig. 10.21, it is inferred that boost-buck converter has low output current ripple of 0.0391 at 40% duty ratio against boost (0.126 at 40% duty ratio) and buck (0.635 at 40% duty ratio) converter. Thus, the boost-buck converter has been chosen for battery charging applications.

10.6 Hardware Results

A laboratory prototype model of the boost-buck converter is developed and tested with low input power. To implement the DC–DC converter semiconductor switches, inductors, capacitors, and resistors are required. With the help of design, suitable

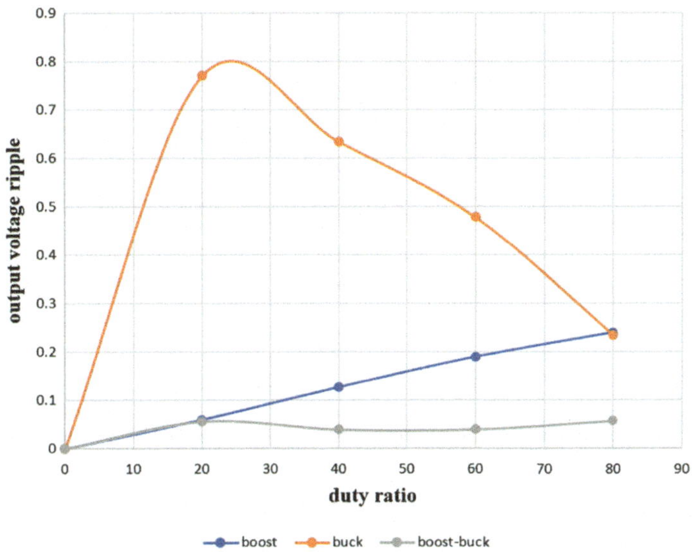

Fig. 10.19 Comparison of output voltage ripple of boost, buck, and boost-buck converter

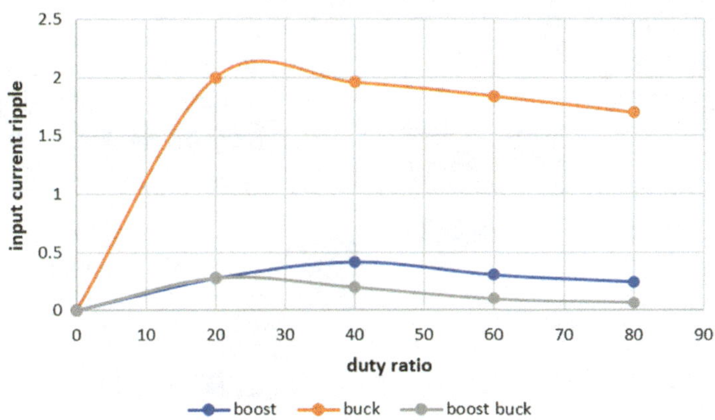

Fig. 10.20 Comparison of input current ripple of boost, buck, and boost-buck converter

inductor and capacitor are selected. The design specifications of the converter are specified in Table 10.1. The power switch is IRFP 250, and the diode is FR 107. Here, MOSFET is used as a switching device. MOSFET is advantageous because of its high switching frequency. Since MOSFET is a voltage-controlled device, to turn on the device a drain–source voltage is applied, which causes the current flow between the gate and source terminal; as a result, the switch is turned on. When the gate pulse is removed, the device turns off. The pulse for the switch is generated using Arduino Uno. From Fig. 10.22, it has been implied that the supply voltage

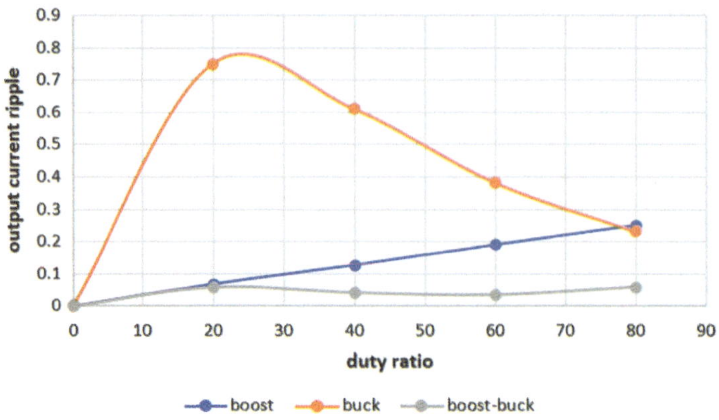

Fig. 10.21 Comparison of output current ripple of boost, buck, and boost-buck converter

Fig. 10.22 Prototype of boost-buck converter

of 9 V is boosted to 10.7 V for $d = 0.16$ and bucked to 10.29 V for $d = 0.95$. The comparison between simulation and hardware results is presented in Table 10.2. Thus, the simulation results are validated.

Table 10.2 Comparison of simulation and hardware results

Voltage	Simulation results (V)	Experimental results (V)
V_1	10.76	10.7
V_0	10.35	10.29

10.7 Conclusion

In this paper, the boost-buck DC–DC converter has been proposed. The performance of this converter such as output voltage ripple (ΔV_0), output current ripple (ΔI_0), and input current ripple (ΔI_i) has been computed and compared with existing topologies. From the analysis, it could be implied that the suggested topology has low output voltage ripple, low output current ripple, and reduced input current ripple. Thus, the efficiency of the converter is improved, and hence, it is a best candidate for battery charging application. The design equation and working principle of the recommended topology have been discussed. A prototype model has been developed and tested. The hardware results are validated through simulation results.

References

1. A.M. Lulhe, T.N. Date, A technology review paper for drives used in electrical vehicle (EV) & hybrid electrical vehicles (HEV), in *International Conference on Control, Instrumentation, Communication and Computational Technologies (ICCICCT)*, Kumaracoil, (2015), pp. 632–636
2. D.M. Bellur, M.K. Kazimierczuk, DC-DC converters for electric vehicle applications, in *Electrical Insulation Conference and Electrical Manufacturing Ex-po, Nashville,* TN (2007), pp. 286–293
3. A.K. Karmaker, S. Roy, M.R. Ahmed, Analysis of the impact of electric vehicle charging station on power quality issues, in *International Conference on Electrical, Computer and Communication Engineering (ECCE)*, Cox'sBazar, Bangladesh (2019), pp. 1–6
4. D. Ranawat, M.P.R. Prasad, A review on electric vehicles with perspective of battery management system, in *International Conference on Electrical, Electronics, Communication, Computer, and Optimization Techniques (ICEECCOT)*, Msyuru, India (2018), pp. 1539–1544
5. T. Mikhail, S. Tatyana, S. Petr, Usage efficiency of renewable energy sources for charging passenger electric transport, in *Renewable Energies, Power Systems & Green Inclusive Economy (REPS-GIE)*, Casablanca (2018), pp. 1–5
6. C. Kanumilli, A. Singh, A. Ganesh, M. Srinivas, Plug in electric solar vehicle, in *Biennial International Conference on Power and Energy Systems: Towards Sustainable Energy (PESTSE)*, Bangalore (2016), pp. 1–4
7. N. Boujelben, F. Masmoudi, M. Djemel, N. Derbel, Design and comparison of quadratic boost and double cascade boost converters with boost converter, in *14th International Multi-Conference on Systems, Signals & Devices (SSD)*, Marrakech (2017), pp. 245–252
8. M.A. Yasko, Analysis, design and simulation of buck converter for photovoltaic system, in *22nd International Conference Electronics*, Palanga, Lithuania (2018), pp. 1–6
9. A.M. Haque, S. Sharma, D. Nagal, Conventional and switched inductor buck boost converter circuit for solar power system: simulation, comparison and results, in *International Conference on Electrical, Electronics, and Optimization Techniques (ICEEOT)*, Chennai (2016), pp. 873–878
10. S. Nahar, M.B. Uddin, Analysis the performance of interleaved boost converter, in *4th International Conference on Electrical Engineering and Information & Communication Technology (iCEEiCT)*, Dhaka, Bangladesh (2018), pp. 547–551
11. A. Ndtoungou, A. Hamadi, A. Missanda, K. Al-Haddad, Modeling and control of a cascaded boost converter for a battery electric vehicle, in *IEEE Electrical Power and Energy Conference*, London, ON (2012), pp. 182–187

12. T.R. Choudhury, B. Nayak, Comparison and analysis of cascaded and quadratic boost converter, in *IEEE Power, Communication and Information Technology Conference (PCITC)*, Bhubaneswar (2015), pp. 78–83
13. M.H. El-Ahmar, A.M. El-Sayed, A.M. Hemeida, Mathematical modeling of photovoltaic module and evaluate the effect of various paramenters on its performance, in *Eighteenth International Middle East Power Systems Conference (MEPCON)*, Cairo (2016), pp. 741–746
14. M.H. Rashid, *Power electronics: circuits, devices and applications*, 2nd edn. (Prentice-Hall Inc., Englewood Cliffs, 1993)
15. C.S. Matwankar, A. Alam, Solar powered closed-loop current controlled dc-dc buck converter for battery charging application, in *International Conference on Vision Towards Emerging Trends in Communication and Networking (ViTECoN)*, Vellore, India (2019), pp. 1–5

Chapter 11
Spoke-Type and Surface-Mounted BLDC Motor for Automotive Applications—A Brief Comparison

Kareti Aasritha, Aneesh Jategaonkar, R. Ruthra Prakashini, and N. C. Lenin

Abstract Permanent magnet brushless DC (PMBLDC) motor has been promoted rapidly in automotive applications. Current automotive vehicle demands more than 40 different types of applications which require rotating motors. This paper compares the surface-mounted and spoke-type PMBLDC motor for the application such as anti-lock braking systems, pedal adjusters, and seat belt pretensioners. Technical specifications, electromagnetic analysis, magnet mass optimization, and cost are discussed in detail. Advantages of both the configurations are compared and concluded finally.

Keywords Brushless DC motor · Automotive application · Surface mounted · Spoke type

11.1 Introduction

Brushless DC motor is becoming more significant in recent times, and they are mainly used in automotive applications [1]. BLDC motors have superior control performance and reliability and are cost effective [2]. Due to the absence of mechanical contact composed of commutator and brushed body, it does not have commutation spark and has a long life and reliable operation [3]. Therefore, it can be used under long-term stable operation load conditions. Length of winding gets reduced due to the construction of poly-phase PMBLDC motor, and as a result copper losses get reduced [4]. BLDC has many advantages over brushed DC motor and induction motor because of the electronic commutation technique and permanent magnet rotor.

The usage of automobiles is getting increased day by day, and more number of motors are being used. Some of the applications of BLDC motor are anti-lock braking

---

K. Aasritha · A. Jategaonkar · R. Ruthra Prakashini
Vellore Institute of Technology, Chennai Campus, Chennai 600127, India

N. C. Lenin (✉)
Electric Vehicles-Incubation, Testing and Research Centre, Vellore Institute of Technology, Chennai, India
e-mail: lenin.nc@vit.ac.in

© The Author(s), under exclusive license to Springer Nature Singapore Pte Ltd. 2021
V. Kamaraj et al. (eds.), *Emerging Solutions for e-Mobility and Smart Grids*, Springer Proceedings in Energy, https://doi.org/10.1007/978-981-16-0719-6_11

system, seat belt pretensioner motor, pedal adjuster, etc. In this paper, the above-mentioned three applications in spoke-type and surface-mounted are designed using finite element method. Section 11.2 describes about BLDC motor, and introduction to the three applications is discussed. Section 11.3 describes the specification and design approach, and Sect. 11.4 describes the results obtained. Finally, it is concluded by discussing the comparison between spoke-type and surface-mounted.

11.2 About BLDC

Brushless DC motor is also known as synchronous motor or BLDC motor. BLDC motor is an electronically commutated motor which does not have brushes in it. BLDC motor has two parts: stator and rotor. Stator is the stationary part which consists of the windings, and rotor is the rotating part; it consists of permanent magnet (in case of BLDC motor). BLDC motor works on the principle that is similar to that of conventional DC motor, i.e., Lorentz law. In this motor, rotor rotates due to the force of attraction between the electromagnet stator and the permanent magnet rotor. This motor is of two types; they are inner rotor motor and outer rotor motor. There is no much difference between both the types, and it is only about the design but they have the same working principle. Figure 11.1 shows the basic structure of PMBLDC motor.

BLDC motors are more efficient due to the presence of permanent magnets, they have high-speed operation in both loaded and unloaded conditions due to the absence of brushes, they have smaller geometry, and they are very light in weight. Due to the absence of brushes, operation is noiseless [5]. The above mentioned are some

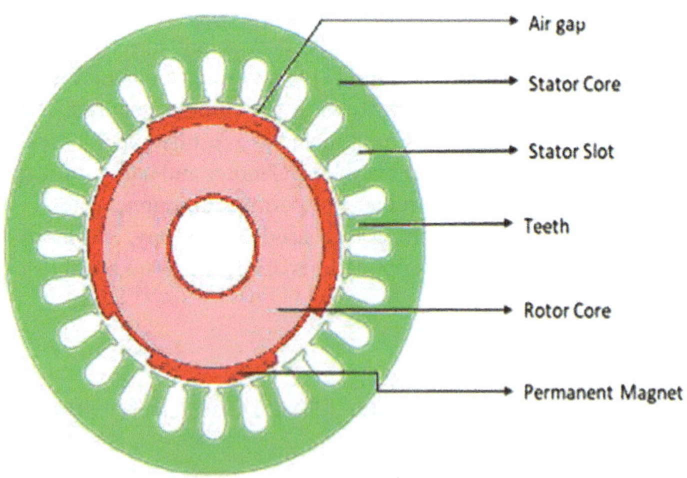

Fig. 11.1 Basic structure of PMBLDC motor

Fig. 11.2 Seat belt pretensioner

advantages of BLDC motor over conventional DC motor. Above mentioned are some of the advantages of BLDC motor over conventional DC motor. This section describes about three applications, namely (i) seat belt pretensioner, (ii) pedal adjuster and (iii) anti-lock braking system.

11.2.1 Seat Belt Pretensioner

Seat belt pretensioner is activated during major accidents like frontal collision, side collision, rollover accidents, and it is not activated for minor accidents. It ensures that the front passenger is not thrown forward and lap pretensioner ensures that the person is not pushed down, i.e., near the dash. Seat belt pretensioner sacrifices its life in an accident, in order to save our life; i.e., it is made only for one-time use, it gets locked once it is activated, and it should be replaced [6]. Seat belt pretensioner is shown in Fig. 11.2.

11.2.2 Pedal Adjuster

Pedal adjusters are used to adjust pedals up to 3 in. closer to driver according to his comfort. These are not only helpful for short-legged people but also to prevent injuries from air bag deployment [7]. Pedal adjuster is shown in Fig. 11.3.

11.2.3 Anti-lock Braking System

Anti-lock braking system is one of the safety equipments that prevents the wheel from getting locked during harsh, panic, and harsh braking conditions. When ABS is activated, brakes are applied and applied numerous times in a second and it ensures

Fig. 11.3 Pedal adjuster [8]

Fig. 11.4 Anti-lock braking system

that the wheel does not get locked [9]. As the vehicle slows down, driver gets grip so that wheel does not skid and helps the driver to have control and braking performance [10]. Structure of anti-lock braking system is shown in Fig. 11.4.

11.3 Specifications and Design Approach

This section describes the specifications of the three applications, and the step-by-step design approach is explained in sequence. Here, both spoke-type and surface-mounted motors are designed for the specifications given below.

In the designed motors, to avoid practical difficulties slot fill factor is always kept less than 45% and the slot shape used for all motors is shown in Fig. 11.5. In all the applications, M36_29G is used in both stator and rotor and ceramic 8D is used as the permanent magnet. Ceramic is very less expensive compared to other magnets, so it is used for designing.

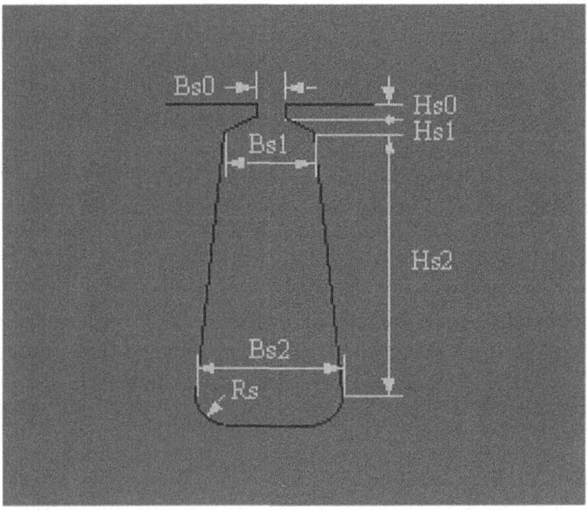

Fig. 11.5 Shape of the stator slot

11.3.1 Seat Belt Pretensioner

The specified specifications for seat belt pretensioner motor are shown in Table 11.1.

$$\theta = T \times \omega \tag{11.1}$$

where

- θ angular displacement of the rotor
- T time in sec
- ω speed in radians per sec

In the above specified motor, it is given that speed is given as 14,500 rpm.

$$\text{Therefore,}\quad \theta = T \times {}^{14{,}500\times 2\times\pi}/_{60} \tag{11.2}$$

Table 11.1 Specifications for seat belt pretensioner

Parameters	Values
Output power	85 W
Voltage	12 V
Speed	14,500 rpm
Torque	0.5 Nm
Stack length	57 mm
Stator outer diameter	35.7 mm
Stator slot fill factor	45%
Shaft diameter	3.175 mm

$$\theta = T \times 1518.437(\text{in radians}) \tag{11.3}$$

$$\theta = T \times 87,000(\text{in degrees}) \tag{11.4}$$

In the above equation, the value of T is varied. By varying the value of T, angular displacement at that particular time instant can be calculated.

(a) **Stator Design**
The designed stator of the motor is configured with 35.7 mm as outer diameter and inner diameter of stator are taken as 19 mm to obtain the perfect results, and it is configured with 12 slots in both spoke-type and surface-mounted model. The dimensions with which slots are designed for seat belt pretensioner motor in spoke-type model are shown in Table 11.2. The designed structure of stator is shown in Fig. 11.6.

(b) **Rotor Design**
The designed rotor of the motor is configured with 18 mm as outer diameter and inner diameter of rotor are taken as 7.2 mm, air gap is specified to be 0.5 mm, to avoid practical difficulties, and it is configured with 10 poles in spoke-type and with 6 poles in surface-mounted. The configured width and thickness of pole are 2.66 mm and 1.47 mm, and shaft diameter is 3.175 mm. The designed structure of rotor is shown in Fig. 11.7.

Table 11.2 Slot dimensions of seat belt pretensioner motor (spoke-type)

Parameters	Values (mm)
Hs0	0.5
Hs1	0.6
Hs2	3.8
Bs0	1.7
Bs1	3.0
Bs2	5.2
Rs	1.3

Fig. 11.6 Stator structure of seat belt pretensioner

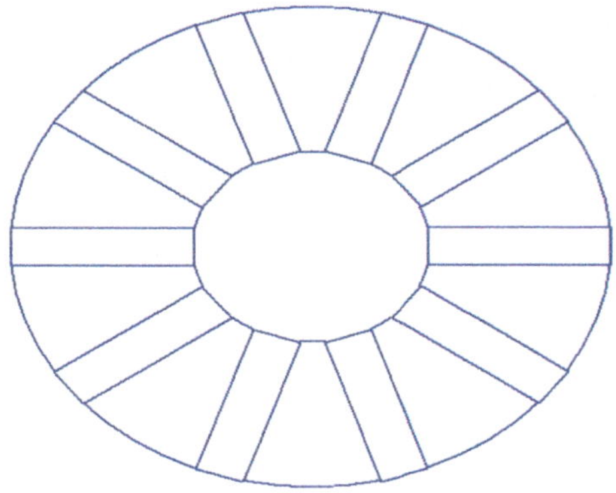

Fig. 11.7 Rotor structure for seat belt pretensioner

(c) **Winding Design**

The designed motor is configured with 14 conductors per slot, and with one strand in spoke-type and in surface-mounted it is configured with 8 conductors per slot and one strand. The wire diameter used in this motor is 0.813 mm. A number of winding layers used in this motor are two, and winding type used is whole-coiled type. Whole-coiled type winding is shown in Fig. 11.8.

(d) **Full Machine**

This motor is operated under constant speed load type. The designed characteristics of this motor are fast response and high starting torque, and circuit type used for this motor is Y-type three-phase winding. Input power of this motor is 106.25 W, losses are specified as 0.53125 W, and the average current drawn is 8.8542 A. Structure of seat belt pretensioner motor is shown in Fig. 11.9.

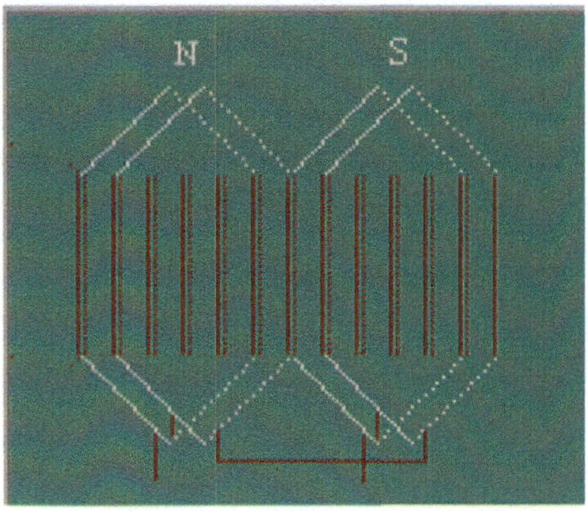

Fig. 11.8 Whole-coiled winding

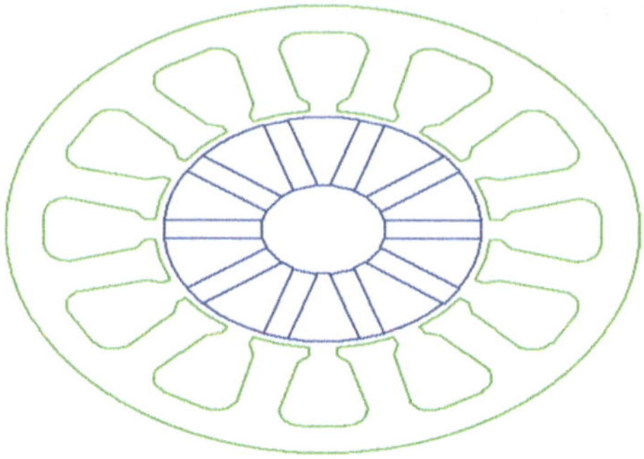

Fig. 11.9 Motor structure of seat belt pretensioner motor (spoke type)

11.3.2 Pedal Adjuster

Necessary characteristics of the motor to make it typically used for pedal adjuster applications that can be easily adapted for other applications are shown in Table 11.3.

(a) **Stator Design**
The designed motor is configured with 39.8 mm as outer diameter, and 18 mm is taken as inner diameter; it is configured with 9 slots in spoke-type model, and it is configured with 12 slots in surface-mounted model. Slots that are designed in order to have higher air gap flux density are shown in Table 11.4. The stator structure of motor is shown in Fig. 11.10.

(b) **Rotor Design**
Rotor of pedal adjuster is designed in such a way that its outer diameter is 17 mm and its inner diameter is 6 mm; it is configured with 6 poles in both spoke-type and surface-mounted. Thickness and width of each pole are 1.8 mm and 3.75 mm, respectively, and shaft diameter is 4.005 mm. The structure of rotor is shown in Fig. 11.11.

(c) **Winding Design**
Two-layered whole-coiled type winding is used in the designed motor. In spoke-type model, motor is configured with 32 conductors per slot and two strands,

Table 11.3 Specifications of pedal adjuster

Parameters	Values
Output power	13.96 W
Voltage	12 V
Speed	3400 rpm
Torque	0.039 Nm
Stack length	72 mm
Stator outer diameter	39.8 mm
Stator slot fill factor	45%
Shaft diameter	4.005 mm

Table 11.4 Slot dimension of pedal adjuster (spoke-type motor)

Parameter	Values (mm)
Hs0	1
Hs1	0.5
Hs2	3.8
Bs0	1
Bs1	5
Bs2	8
Rs	1

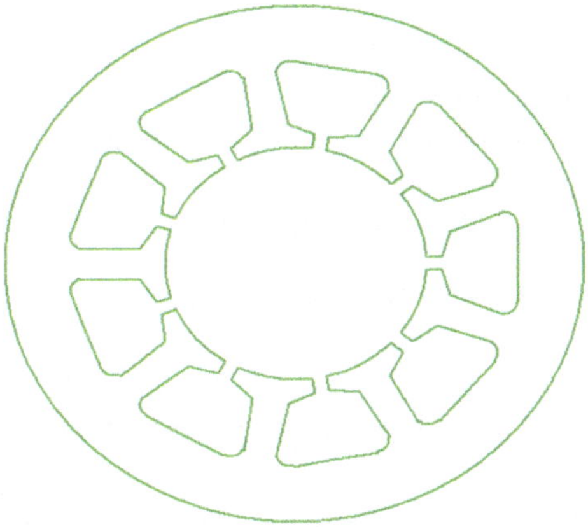

Fig. 11.10 Stator structure of pedal adjuster motor

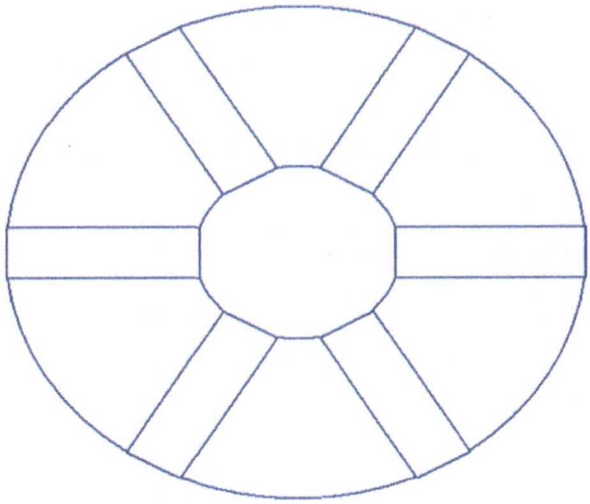

Fig. 11.11 Rotor structure for pedal adjuster

whereas in surface-mounted model it is configured with 24 conductors per slot and one strand is used. The average current to be drawn is 1.45 A, and the required diameter of wire is 0.455 mm.

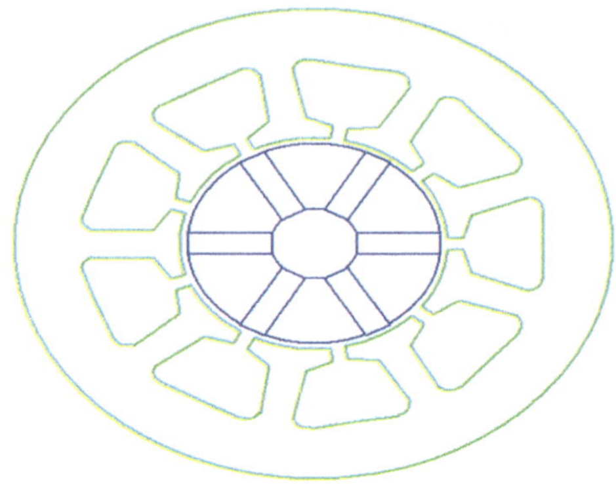

Fig. 11.12 Fully designed pedal adjuster motor (spoke type model)

(d) **Full Machine**
Star-type circuit is used in this motor, and it is operated at constant speed load type. Theoretical value of input power for this motor is 17.45 W. This motor is not only an application of pedal adjuster but can be easily adapted for other applications. Air gap is 0.5 mm, and losses specified for this motor are 0.0873 W. Fully designed pedal adjuster motor is shown in Fig. 11.12.

11.3.3 Anti-lock Braking System

Required design data for anti-lock braking system motor is shown in Table 11.5.

(a) **Stator Design**
The configured outer and inner diameters of the designed stator are 46 mm and 22 mm, respectively. The motor is configured with 9 slots in spoke-type and with 12 slots in surface-mounted type. Slot dimensions of the motor are shown in Table 11.6, and the structure of stator is shown in Fig. 11.13.

Table 11.5 Design data for anti-lock braking system

Parameters	Values
Output power	53 W
Voltage	12 V
Speed	6100 rpm
Torque	0.082 Nm
Stack length	23 mm
Stator outer diameter	46 mm
Stator slot fill factor	45%
Shaft diameter	4.000 mm

Table 11.6 Slot dimension of anti-lock braking system (spoke-type motor)

Parameter	Values (mm)
Hs0	0.7
Hs1	0.7
Hs2	5
Bs0	1.4
Bs1	7.2
Bs2	10.2
Rs	1.8

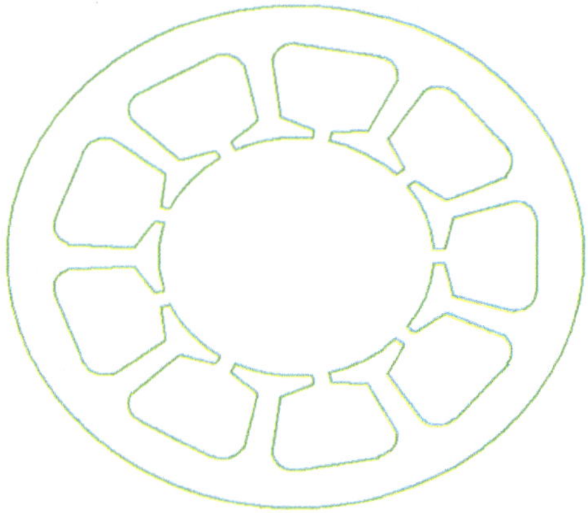

Fig. 11.13 Stator structure of anti-lock braking system

(b) **Rotor Design**
The air gap of this motor is 0.5 mm; therefore, outer diameter is 21 mm and inner diameter is 4.4 mm. Specified shaft diameter is 4.000 mm, and the motor is configured with 6 poles in both spoke-type and surface-mounted models. Thickness and width of each pole are 1.8 mm and 7.64 mm, respectively. The structure of rotor is shown in Fig. 11.14.

(c) **Winding Design**
Motor is a two-layered, whole-coiled winding type. A number of conductors per slot are 26 and 22 in spoke-type and surface-mounted type, respectively. A number of strands used in spoke-type and surface-mounted are two and one, respectively. The diameter of the wire used in this motor is 0.724 mm.

(d) **Full Machine**
Input power required for the designed motor is 66.25 W, and the average current drawn is 5.52 A. This motor's operation is at constant speed load type, the circuit type used is star type. This motor is operated at constant speed load type and star type circuit is used. Loss specified for the motor is 0.33125 W. The structure of anti-lock braking system is shown in Fig. 11.15.

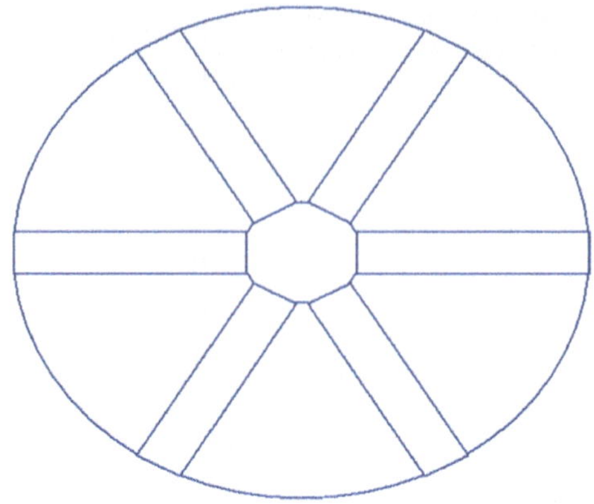

Fig. 11.14 Rotor structure for anti-lock braking system

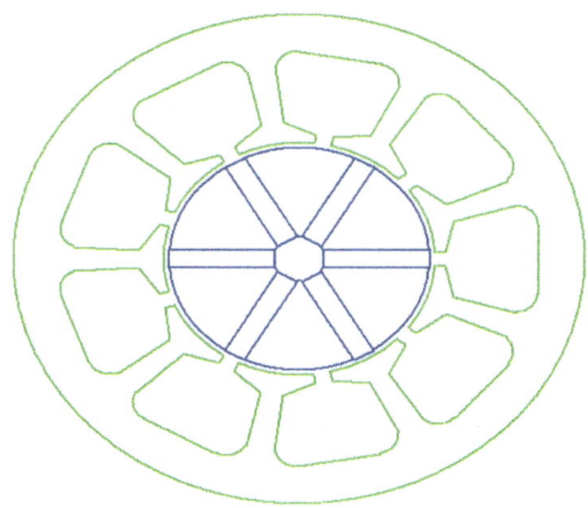

Fig. 11.15 Motor structure of anti-lock braking system motor (spoke-type)

11.4 Results and Discussion

Before manufacturing, performance estimation is crucial. Effectiveness of a motor can be judged by looking at critical parameters like speed, torque, core and copper loss, output power, efficiency, etc. In a BLDC motor, input voltage is directly proportional to speed, efficiency gradually increases with speed and decreases after maximum efficiency is reached, and torque and speed are inversely proportional; this section describes how parameters like output power, efficiency, output torque, input current vary with speed in all the three applications.

11.4.1 Seat Belt Pretensioner

Analytical results of seat belt pretensioner are shown in Table 11.7, and the important characteristic plots are shown in Figs. 11.16, 11.17, 11.18 and 11.19. Stator slot

Table 11.7 Results for seat belt pretensioner

Parameters	Values	
	Spoke-type	Surface-mounted
Average input current (A)	10.1614	9.60346
Root-mean-square armature current (A)	12.5974	9.6028
Frictional and windage loss (W)	1.0625	0.85
Iron core loss (W)	1.63701	2.2443
Armature copper loss (W)	17.2312	12.68
Transistor loss (W)	13.3115	10.7414
Diode loss (W)	3.31306	1.3355
Total loss (W)	36.5552	27.8509
Output power (W)	85.3817	87.3906
Input power (W)	121.937	115.241
Efficiency (%)	70.0212	75.8325
Rated speed (rpm)	14,500	14,500
Rated torque (Nm)	0.05623	0.057553
Stator slot fill factor (%)	42.8116	43.9893

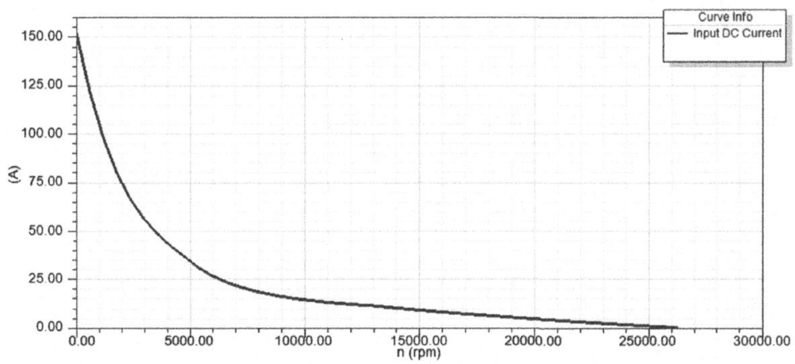

Fig. 11.16 Input DC current versus speed graph (spoke-type)

fill factor is obtained as 42.8 and 44% in spoke-type and surface-mounted motor, respectively, which makes windings physically easy.

From the above figures, we can say that at rated speed (i.e., at 14,500 rpm) input DC is approximately 10 A, the efficiency of the motor at 5000 rpm is 25%, at rated speed it is approximately 70% and at 20,000 rpm efficiency is high, i.e., approximately 77.5%, beyond that efficiency got gradually decreased and reached zero beyond 26,000 rpm. Output power is recorded to be 103 W at 5000 rpm, approximately 85 W at rated speed and approximately 48 W at 20,000 rpm. At rated speed, output torque is noted to be 0.05 Nm.

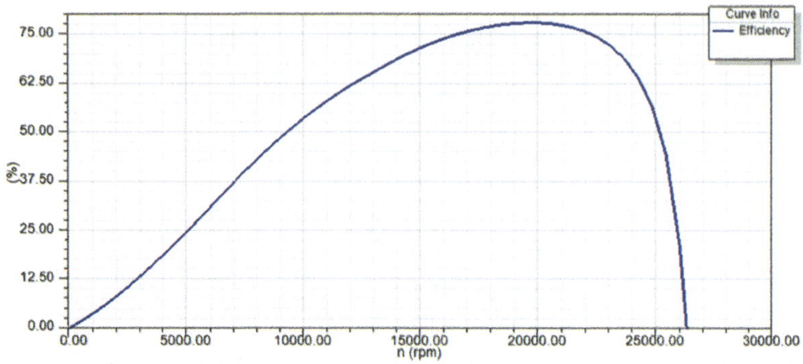

Fig. 11.17 Efficiency versus speed graph (spoke-type)

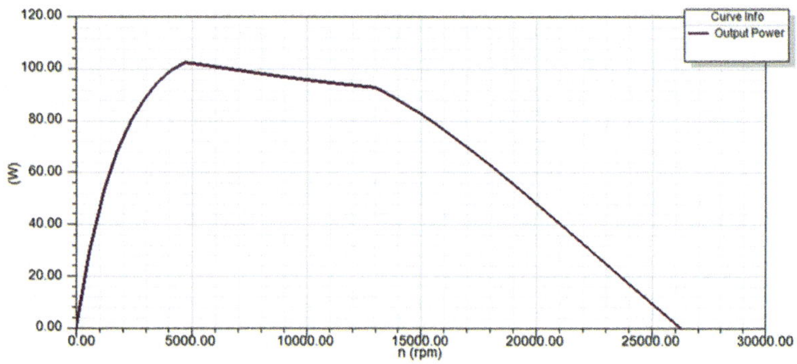

Fig. 11.18 Output power versus speed graph (spoke-type)

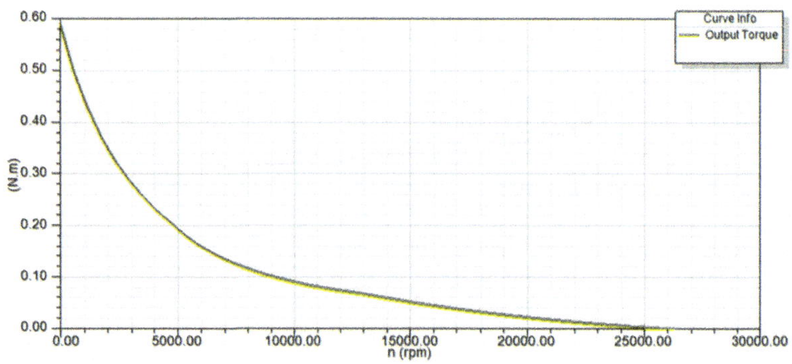

Fig. 11.19 Output torque versus speed graph (spoke-type)

11.4.2 Pedal Adjuster

Table 11.8 shows the characteristic results of pedal adjuster application. Figures 11.20, 11.21, 11.22, 11.23 and 11.24 show the important characteristic plots of pedal adjuster motor applications. The obtained stator slot fill factors in surface-mounted and spoke-type are 45.65 and 41%, respectively.

From the above plots, it is observed that input DC at rated speed (i.e., 3362.47 rpm) is approximately 1.6 A. Efficiency of the motor at 1000 rpm seems to be 20%; at rated

Table 11.8 Results for pedal adjuster motor

Parameters	Values	
	Spoke-type	Surface-mounted
Average input current (A)	1.71	1.85837
Root-mean-square armature current (A)	1.62	1.6439
Frictional and windage loss (W)	0.17	1.396
Iron core loss (W)	0.41	0.6855
Armature copper loss (W)	3.94	3.13954
Transistor loss (W)	1.85	1.90576
Diode loss (W)	0.18	0.104314
Total loss (W)	6.55	7.23111
Output power (W)	13.96	15.0694
Input power (W)	20.51	22.3005
Efficiency (%)	68.06	67.5742
Rated speed (rpm)	3362.47	3400
Rated torque (Nm)	0.04	0.042324
Stator slot fill factor (%)	41.058	45.6288

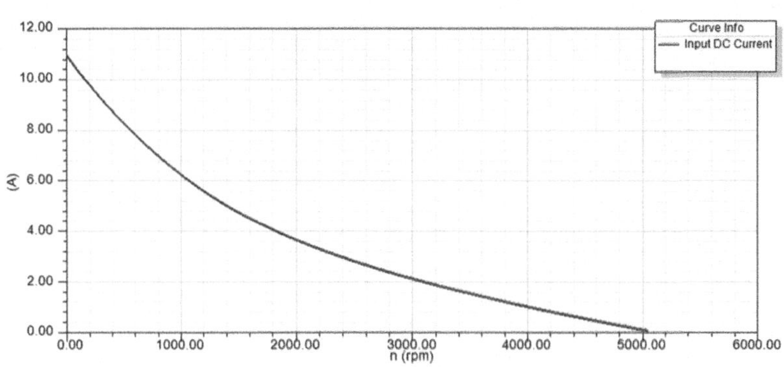

Fig. 11.20 Input DC current versus speed graph (spoke-type)

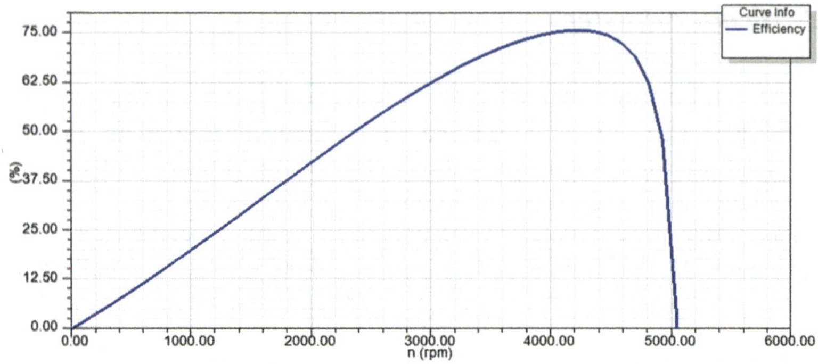

Fig. 11.21 Efficiency versus speed graph (spoke-type)

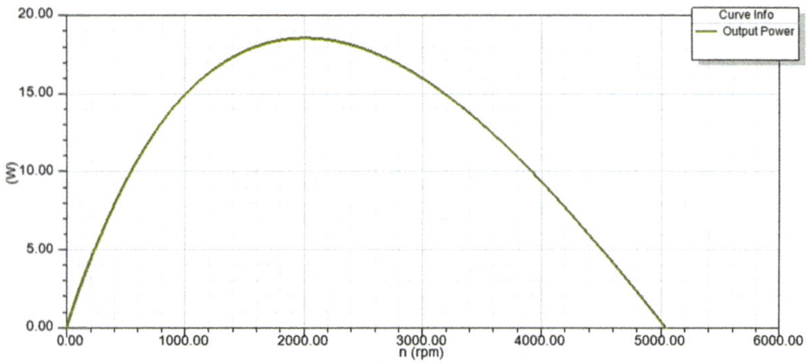

Fig. 11.22 Output power versus speed graph (spoke-type)

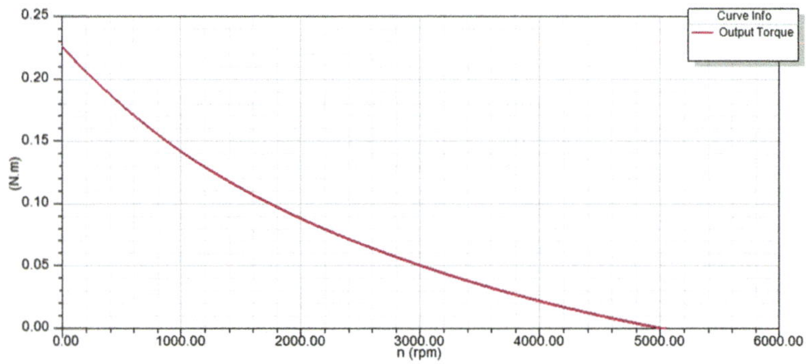

Fig. 11.23 Output torque versus speed graph (spoke-type)

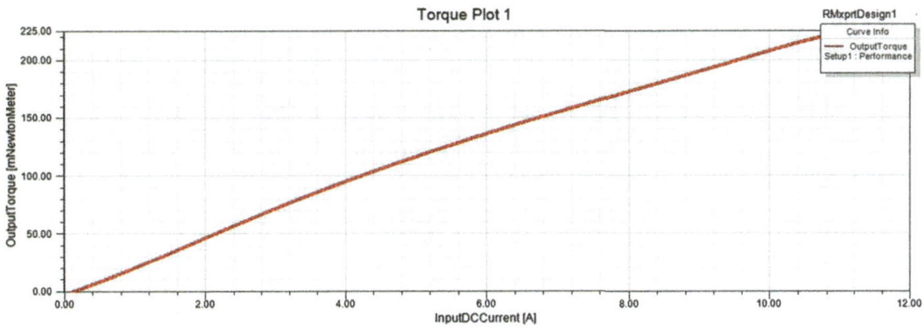

Fig. 11.24 Input current versus output torque (spoke-type)

speed, it is approximately 67.5% and maximum efficiency is obtained at 4000 rpm, i.e., 75%, beyond which efficiency started to decrease and reached zero at 5000 rpm. Output power is observed to be 15 W at 1000 rpm, at rated speed it is approximately 13 W, and at 4500 rpm it is observed to be 5 W. The output torque of the motor at rated speed is observed to be 0.04 Nm.

11.4.3 Anti-lock Braking System

Stator slot fill factors of anti-lock braking system in spoke-type model and surface-mounted type are 45% and 44.67%, respectively. Results of this motor are shown in Table 11.9, and important characteristic plots are shown in Figs. 11.25, 11.26, 11.27, 11.28 and 11.29.

As shown in above plots, the input current at rated speed (i.e., 6100 rpm) is approximately 7 A, efficiency at 2000 r pm is 12.5%, at rated speed it is approximately 63%, at 8000 rpm it is 75%, and efficiency reaches zero at 11200 rpm. The obtained output torque at rated speed is 0.085 Nm, output power at 2000 rpm is 42.5 W, at 6100 rpm it is approximately 52.5 W, and torque becomes zero at 11,200 rpm.

11.5 Comparison of Weights Between Spoke-Type and Surface-Mounted Motors

As mentioned in the above sections, motors were designed in spoke-type model and surface-mounted model and required output power and efficiency were also obtained. Now, this section describes the amount of material consumed and shows the bar graphs of some of the important materials.

Table 11.9 Results for anti-lock braking system

Parameters	Values	
	Spoke-type	Surface-mounted
Average input current (A)	6.925	6.747
Root-mean-square armature current (A)	9.161	6.862
Frictional and windage loss (W)	0.6625	0.85
Iron core loss (W)	1.238	16.625
Armature copper loss (W)	15.868	16.625
Transistor loss (W)	9.524	7.664
Diode loss (W)	2.725	0.998
Total loss (W)	30.017	27.0328
Output power (W)	53.080	56.9362
Input power (W)	83.097	80.969
Efficiency (%)	63.877	66.6133
Rated speed (rpm)	6100	6100
Rated torque (Nm)	0.0831	0.08443
Stator slot fill factor (%)	45.076	44.6684

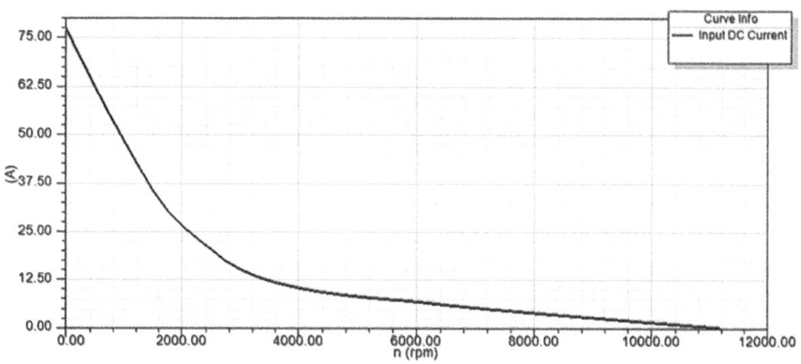

Fig. 11.25 Input DC current versus speed graph (spoke-type)

11.5.1 Seat Belt Pretensioner

Material consumed is calculated, and ferrite magnets are ten times cheaper than other magnetic materials, so ferrite magnets are preferred. Comparison of material consumed is shown in Table 11.10, and comparison of weights using bar graph is shown in Fig. 11.30. Weight of armature core steel is high in spoke-type compared to surface-mounted, and the remaining weights are less in spoke-type.

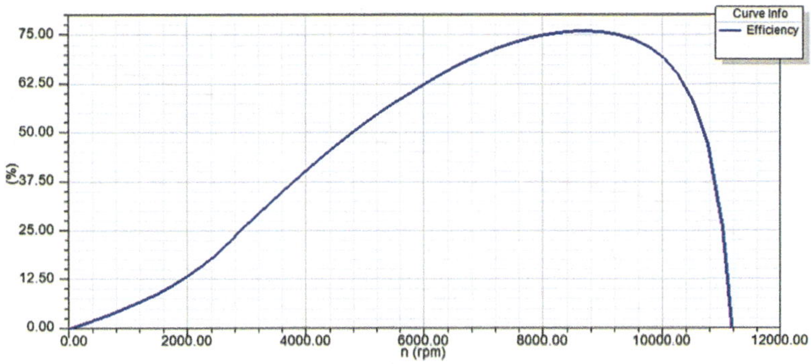

Fig. 11.26 Efficiency versus speed graph (spoke-type)

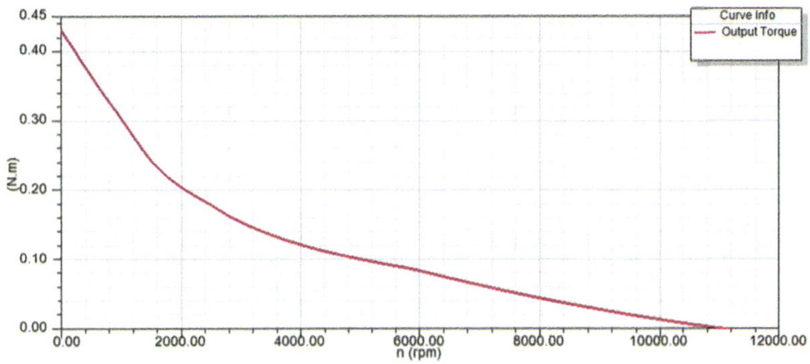

Fig. 11.27 Output torque versus speed graph (spoke-type)

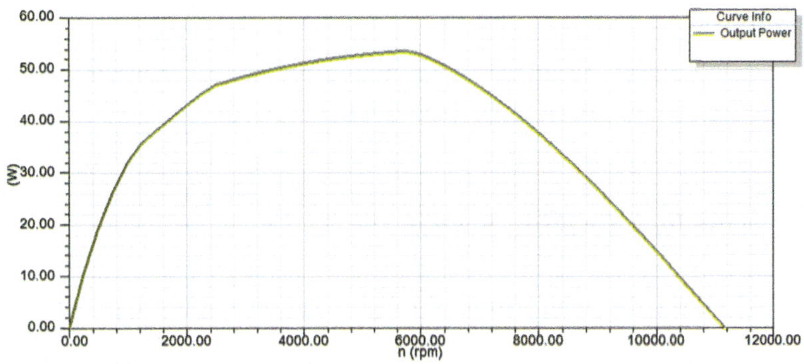

Fig. 11.28 Output power versus speed graph (spoke-type)

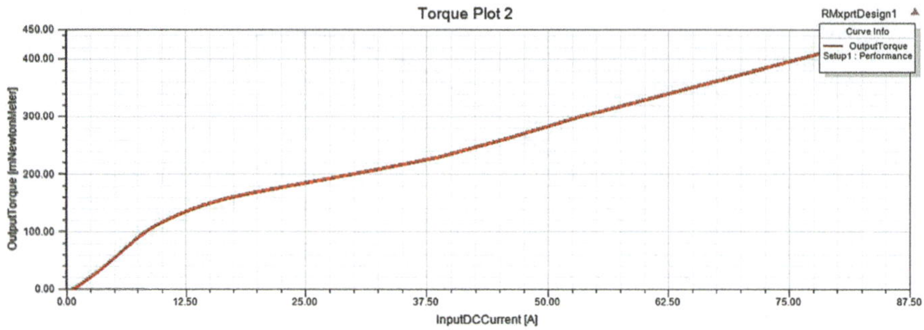

Fig. 11.29 Input current versus torque graph (spoke-type)

Table 11.10 Amount of materials consumed and weight of materials of seat belt pretensioner

Parameters	Values	
	Spoke-type	Surface-mounted
Armature copper weight (kg)	0.048005	0.048211
Permanent magnet weight (kg)	0.010921	0.012317
Armature core steel weight (kg)	0.17969	0.160326
Rotor core steel weight (kg)	0.056028	0.102133
Total net weight (kg)	0.294644	0.322987

11.5.2 Pedal Adjuster

Comparison of weights and amount of materials consumed are shown in Table 11.11. Comparison of weights in bar graph is shown in Fig. 11.31.

11.5.3 Anti-lock Braking System

Table 11.12 shows the comparison of weights between spoke-type and surface-mounted. The weights of permanent magnet in spoke-type and surface-mounted for this application are 9.3 g and 10.8 g, respectively, and it can be said that the weight of permanent has been reduced in spoke-type compared to surface mounted. Comparison of weights using bar graph is shown in Fig. 11.32.

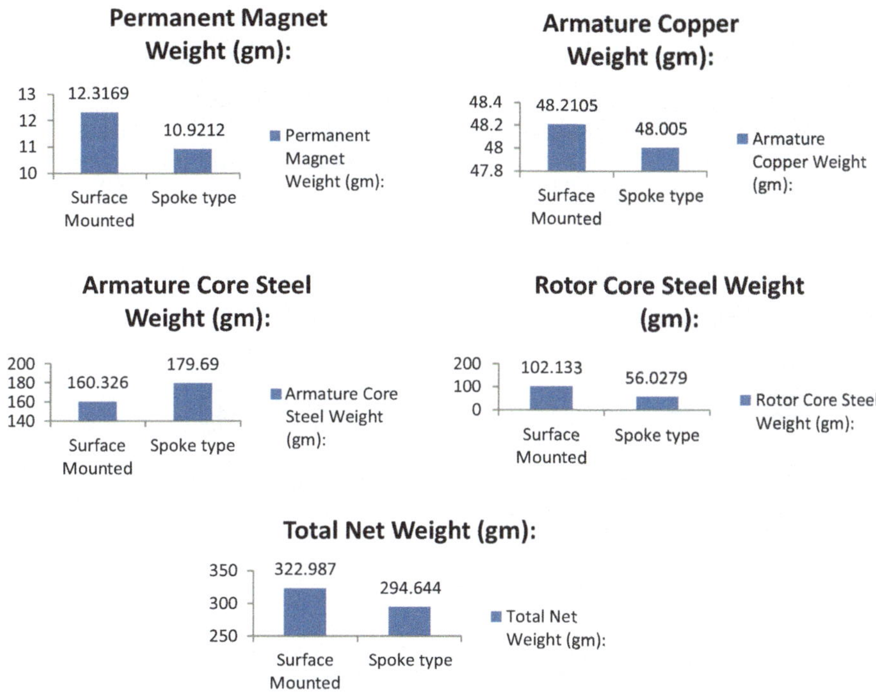

Fig. 11.30 Comparison of weights using bar graphs

Table 11.11 Amount of materials consumed and weight of materials of pedal adjuster

Parameters	Values	
	Spoke-type	Surface-mounted
Armature copper weight (kg)	0.0653	0.0808
Permanent magnet weight (kg)	0.0143	0.0163
Armature core steel weight (kg)	0.3564	0.2739
Rotor core steel weight (kg)	0.0734	0.1418
Total net weight (kg)	0.5093	0.5128

11.6 Conclusion and Future Scope

In this paper, comparison of some of the applications like seat belt pretensioner, pedal adjuster, and anti-lock braking system in two pole types, i.e., in surface-mounted and spoke-type, is done. From the analysis, it can be concluded that both the types have its own advantages. For a spoke-type machine, main drawback is the torque production

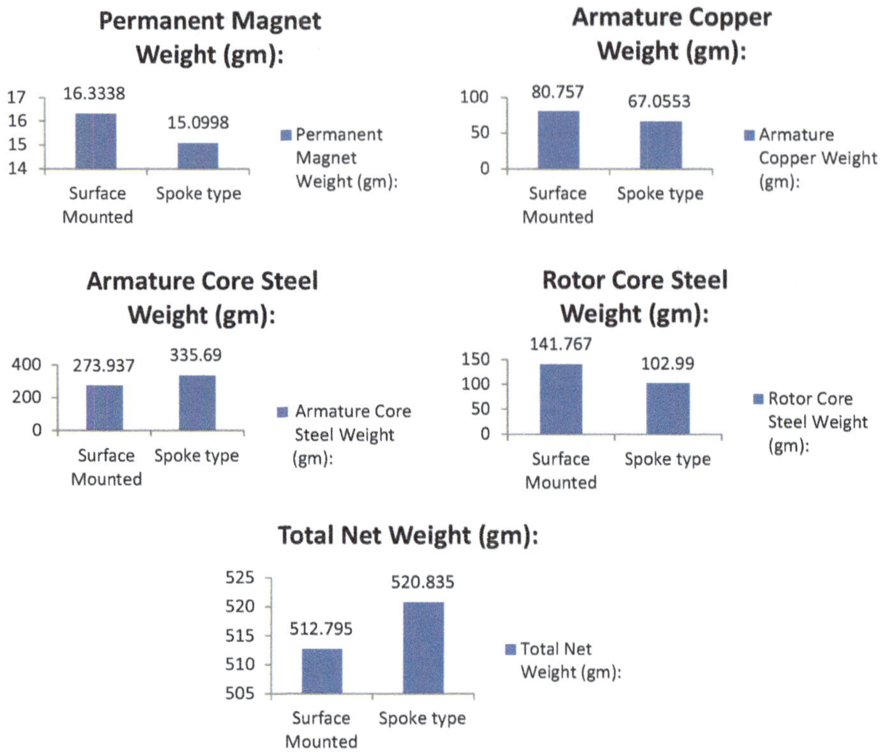

Fig. 11.31 Comparison of weights using bar graphs

Table 11.12 Amount of materials consumed and weight of materials of anti-lock braking system

	Values	
Parameters	Spoke type	Surface mounted
Armature copper weight (kg)	0.052569	0.030913
Permanent magnet weight (kg)	0.009299	0.010811
Armature core steel weight (kg)	0.118047	0.099033
Rotor core steel weight (kg)	0.040634	0.086646
Total net weight (kg)	0.220548	0.227402

is less compared to surface mounted, weight of the magnet is reduced in spoke-type machine, and the variation can be seen in the above comparison chart. As the weight of the permanent magnet reduces, efficiency reduces. Since the weight of the permanent magnet in spoke-type motor is less, efficiency of spoke-type motor is less compared to surface-mounted motor. Though torque produced is less, other characteristics are almost similar. The future work is to increase the torque produced in spoke-type motor and try to reduce the magnet weight in surface-mounted motor.

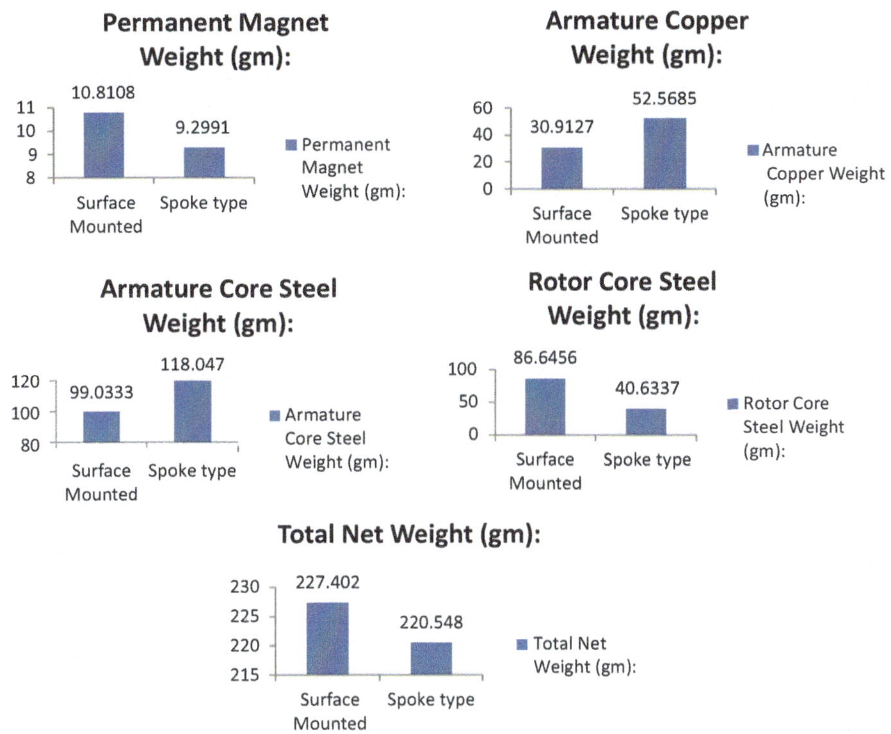

Fig. 11.32 Comparison of weights using bar graphs

References

1. E Elakkai, in *Design and Modelling of BLDC Motor for Automotive Applications* (2015)
2. Brushless DC Motors: Low Inertia, Fast response. http://www.ohioelectricmotors.com
3. M.M. Rahman, K. Kim, J. Hur, Design and analysis of neodymium free SPOKE-type motor with segmented wing shape permanent-magnet for concentrating flux density, in *2013 IEEE Energy Conversion Congress and Exposition*, Denver, CO (2013), pp. 4991–4997. https://doi.org/10.1109/ecce.2013.6647374
4. K.L. Shenoy, M.S. Kumar, Design topology and electromagnetic field analysis of Permanent Magnet Brushless DC motor for electric scooter application, in *2016 International Conference on Electrical, Electronics, and Optimization Techniques (ICEEOT)*, Chennai (2016), pp. 1541–1545. https://doi.org/10.1109/iceeot.2016.7754942
5. A.J. Pawar, A. Patil, Design and development of 48 V PMBLDC motor for radiator fan application by using ANSYS Maxwell software, in *2017 Third International Conference on Sensing, Signal Processing and Security (ICSSS)* (2017), pp. 247–252. https://doi.org/10.1109/ssps.2017.8071600
6. HowStuffWorks.com. https://auto.howstuffworks.com/car-driving-safety/safety-regulatory-devices/seatbelt-pretensioner.htm 6 July 2020
7. "Ford's Pedal Adjusters Prompt Legal Dispute" 18 Feb 2002. Los Angeles Times www.latimes.com/archives/la-xpm-2002-feb-18-fi-preview18-story
8. DES Gasoling mobility ltd. https://www.desgoslingmobilityltd.co.uk/products/driving-controls/pedal-extensions/

9. G. Genta, L. Morello, Prime movers for motor vehicles, in *The Automotive Chassis*. Mechanical Engineering Series (Springer, Cham, 2020)
10. K.W.E. Cheng, Recent development on electric vehicles, in *2009 3rd International Conference on Power Electronics Systems and Applications, PESA 2009* (2009), pp. 1–5

Chapter 12
Electromagnetic Performance and Thermal Analysis of Reluctance Synchronous and Magnet-Assisted Reluctance Synchronous Motors for Industrial Pump Application

V. S. Nagarajan, M. Balaji, V. Kamaraj, K. Subash, and Rahul Tiwari

Abstract This work is intended to study and compare the performance of 15 kW three-phase reluctance synchronous motor, ferrite-assisted reluctance synchronous motor along with induction motor for centrifugal pump application used in fluid flow application of food and process industries in terms of torque density, efficiency, volume and thermal behavior. The electromagnetic and thermal behavior is predicted by MagNet and ThermNet FEA software. The work involves an initial prototyping and validation of predicted performance by means of a motor–pump system coupled to tank setup. Also, this procedure is initiated through establishment of operating points by matching the speed–torque characteristics of motors and pump system load.

Keywords MagNet · Pump · ThermNet · Torque · RelSyn motor · Ferrite-assisted RelSyn motor

12.1 Introduction

Pumps are used in movement of fluids [1] from one region to another against the gravity in industries and domestic applications as well as in long-distance transportation of essential liquids such as water and gasoline.

The electric motor-driven pumps originated in early part of the twentieth century with improved efficiency compared to steam-driven pumps and had ease of implementation due to increased availability of electricity.

Further, the efficiency of the electric motor-driven pumps was improved by employing variable frequency drives in recent times, which led to increased energy saving, improved reliability, longevity and wide speed range of operation.

Traditionally, induction motors are used in industrial pump applications owing to its rugged construction, self-starting capability and ease of service.

V. S. Nagarajan (✉) · M. Balaji · V. Kamaraj · K. Subash · R. Tiwari
EEE Department, SSN College of Engineering, OMR, Kalavakkam 603110, India
e-mail: nagarajanvs@ssn.edu.in

© The Author(s), under exclusive license to Springer Nature Singapore Pte Ltd. 2021

V. Kamaraj et al. (eds.), *Emerging Solutions for e-Mobility and Smart Grids*, Springer Proceedings in Energy, https://doi.org/10.1007/978-981-16-0719-6_12

Though induction motors have the abovementioned advantages, the power density is less when it is designed to meet out high efficiency as per international efficiency (IE) standards. Also, there is a requirement of forced cooling for high power rating.

Therefore, this work explores the suitability of reluctance synchronous and ferrite magnet-assisted reluctance synchronous motors [2–4] for a 15 kW industrial pump application, which have gained popularity in recent times for their unique performance features. The comparison among the motors is made in terms of efficiency, volume, torque density and thermal performance.

The paper contains the following sections: Sect. 12.2 deals with the overview of centrifugal pumps employed in industries. Section 12.3 describes the key characteristics related to pumps. Section 12.4 details the operating point establishment with respect to speed–torque characteristics with a prototype motor. In Sect. 12.5, comparison of performance measures and thermal characteristics is presented. Section 12.6 describes the conclusion.

12.2 Overview of Centrifugal Pumps

Centrifugal pumps are widely used pump for fluid transportation. A typical centrifugal pump along with its parts is represented in Fig. 12.1. The impeller and volute casing of the pump play a crucial role in the operation of the centrifugal pump. The fluid which is to be pumped enters the eye of the impeller, and the rotating impeller (usually rotated by an electric motor) imparts energy to the fluid due to which the velocity of the fluid is increased. The fluid then enters the volute casing where the velocity drops and the pressure of the fluid is raised. Thus, the fluid coming out has an increased pressure. These pumps are designed for transportation of low viscous liquids like water.

Centrifugal pumps are generally classified as shown in Fig. 12.2. Net positive suction head [1] is an important term in a pumping system as it is related to the

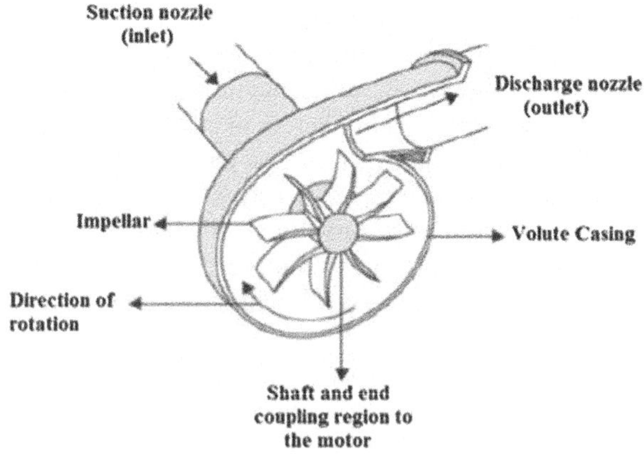

Fig. 12.1 Parts of a centrifugal pump [5]

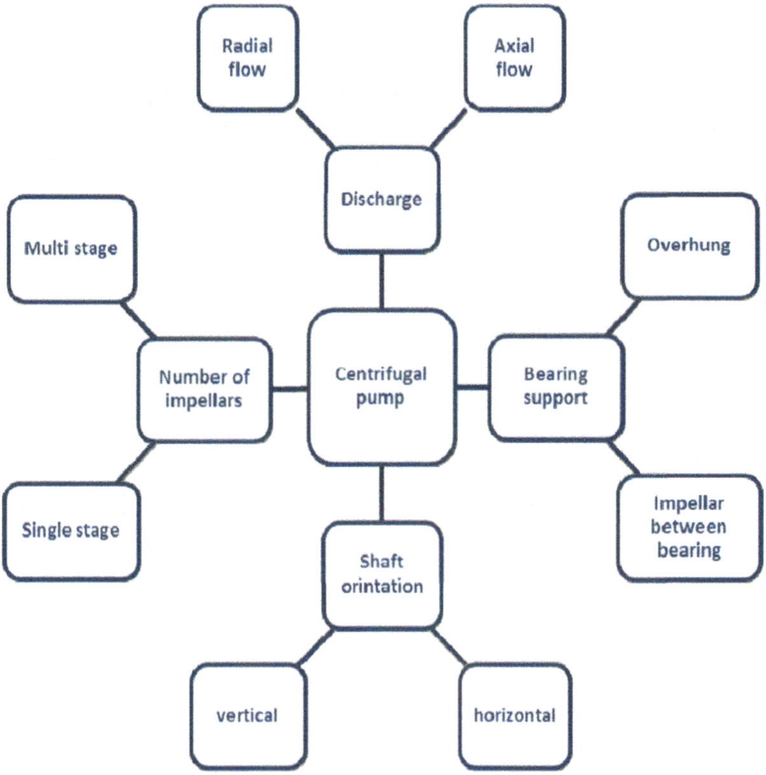

Fig. 12.2 Classification of centrifugal pumps

longevity of the pump. Net positive suction head available (NPSHA), which is defined as the absolute pressure available at the suction side of pump, should be greater than net positive suction head required (NPSHr), which is defined as the minimum pressure required at the suction side in order to prevent cavitation, which is a phenomenon where bubbles or cavity formed in the fluid hits the impeller with high velocity degrading the impeller slowly.

12.3 Pump Performance Characteristics

Operation of centrifugal pumps is governed by affinity laws, which are used to express the relationship between variables involved in pump performance and power.

12.3.1 Affinity Laws

The affinity laws [1] are helpful in the prediction of the performance of centrifugal pump when the diameter of the impeller or speed of the pump is changed.

The affinity laws for centrifugal pump with constant diameter are represented in Figs. 12.3, 12.4 and 12.5. They are defined as

- Flow or volume varies linearly with speed as shown in Fig. 12.3.
- Pressure or head varies proportionally as a square of the speed as shown in Fig. 12.4.
- Power or energy consumption varies proportionally as the cube of speed as shown in Fig. 12.5.

Fig. 12.3 Flow versus speed

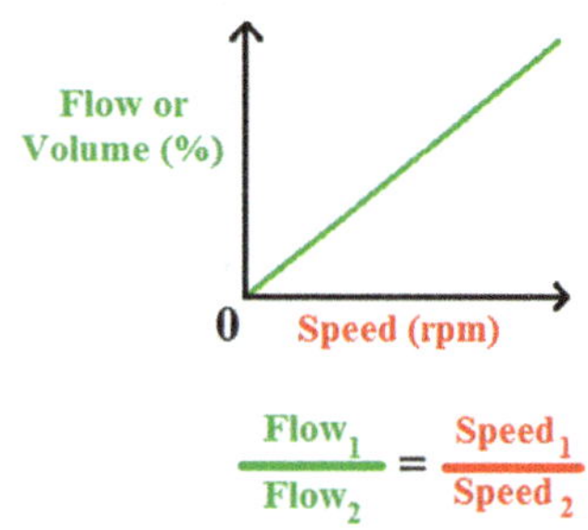

Fig. 12.4 Head versus speed

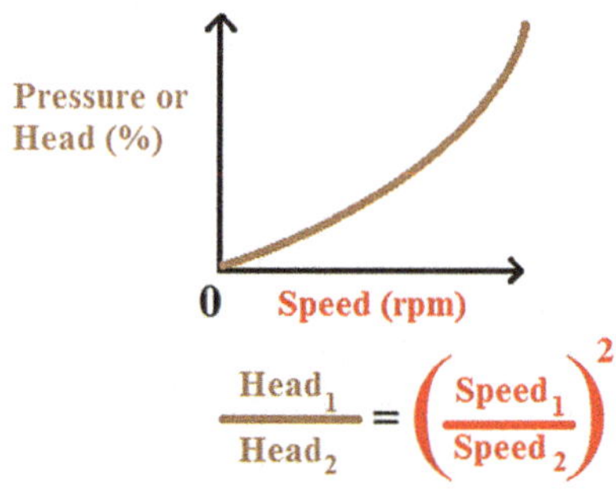

Fig. 12.5 Power versus speed

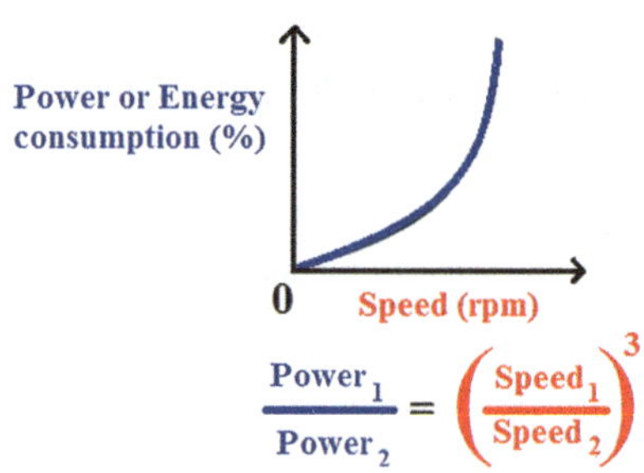

By the understanding and effective utilization of affinity laws, it can be show that pumps have a higher potential for energy savings. For an example, as the speed requirement of the pump decreases by one-half, referring to Fig. 12.5, the power requirement drops down to one-eighth. Therefore, this methodology of utilization of affinity laws is effectively achieved by variable frequency drives which enable energy saving in the motor–pump applications.

12.3.2 System Characteristics

Effective fluid movement along a pipe to a destination depends on static and friction head present in the system [1]. The characteristics providing the relationship between head and flow as shown in Fig. 12.6 are called as system characteristics, which provides information about head that the pump must overcome in order to maintain a constant flow.

In the above curve, static head is the difference in height between the source and destination in a pumping system whereas friction or dynamic head is defined as the frictional loss encountered by the fluid when it moves along in the pipes, valves and other piping systems. Friction head varies with square of the flow rate (which is directly proportional to speed of the motor driving the pump).

12.4 Test Bench of Prototype Pump–Motor System

The test bench shown in Fig. 12.7 consists of 1.1 kW [2] reluctance synchronous prototype motor coupled to centrifugal pump. The pump system connected to a

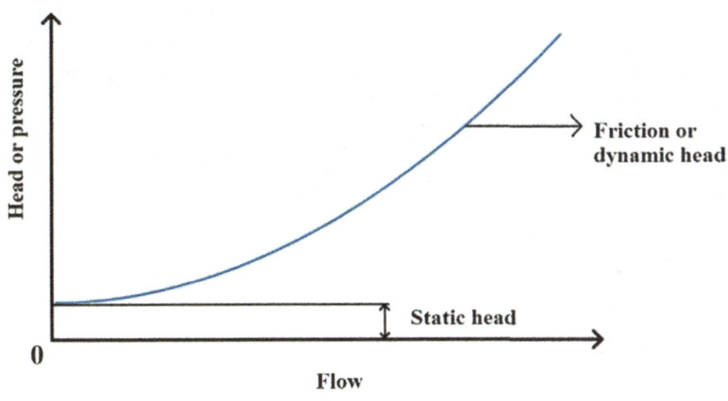

Fig. 12.6 System curve

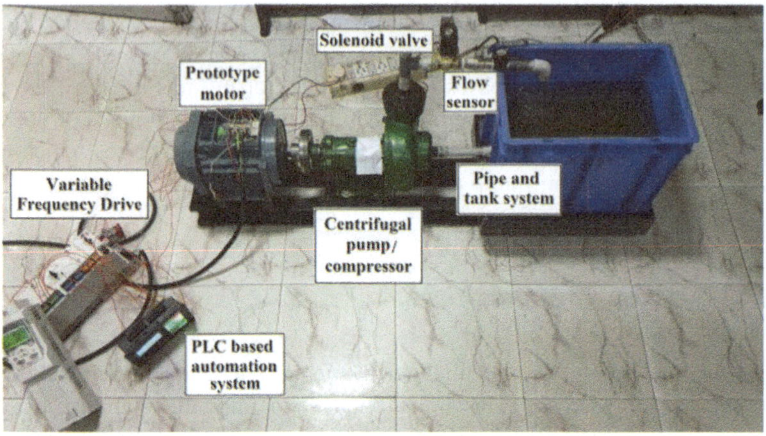

Fig. 12.7 Pump–motor test bench

tank is a closed-loop circulating system. The speed–torque characteristics of reluctance synchronous motor along with the system curve for the closed-loop circulating system which exhibits only frictional loss are shown in Fig. 12.8.

From the above characteristic curves, the operating points are established for operation at a different head or torque with corresponding flow or speed. The reluctance synchronous motor is modeled using FEA, and analysis is carried out for various speeds. The torque performance results also obtained through analytical method are presented in Table 12.1.

From Table 12.1, it can be observed that the torque obtained from FEA is in agreement with the results from the test bench. Hence, further analysis for a 15 kW motor design involves FEA.

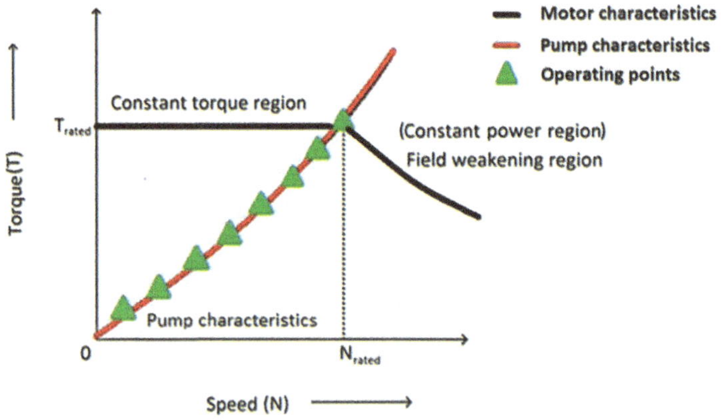

Fig. 12.8 Speed–torque characteristics of reluctance synchronous motor

Table 12.1 Comparison of average torque obtained by FEA, analytical and experimental results

Speed (rpm)	FEA result (Nm)	Experimental result (Nm)
1500	7.17	7.08
1000	4.39	4.25
300	1.30	1.19

12.5 Performance Comparison of 15 kW Motor

Performance analysis is performed for 15 kW induction motor, reluctance synchronous motor and ferrite-assisted reluctance synchronous motor using MagNet and ThermNet [6]. The procedure involved in the design is represented in Fig. 12.9.

12.5.1 Electromagnetic Analysis

Quarter model of reluctance synchronous motor and ferrite-assisted reluctance synchronous motor selected for design is shown in Fig. 12.10.

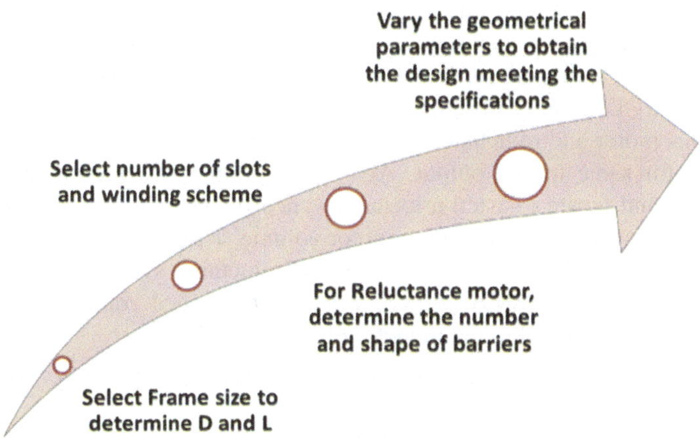

Fig. 12.9 Design procedure [7]

Fig. 12.10 1/4th model of reluctance synchronous motor and ferrite-assisted reluctance synchronous motor

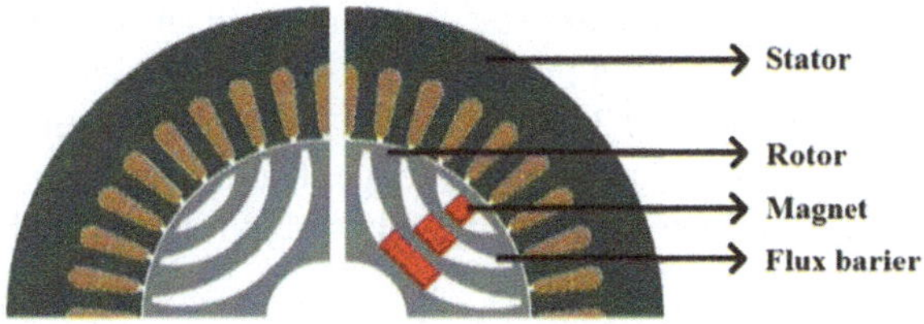

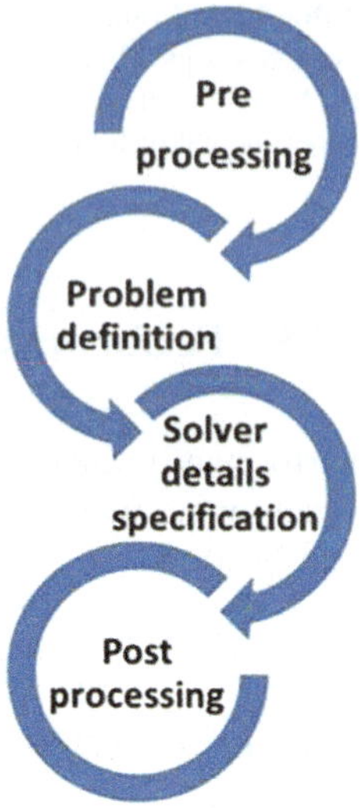

Fig. 12.11 Electromagnetic FEA procedure

The steps involved in electromagnetic FEA are represented in Figs. 12.11 and 12.12.

The electromagnetic analysis results are presented in Table 12.2. The models of induction, reluctance synchronous and ferrite-assisted reluctance synchronous motors for 15 kW rating are represented in Figs. 12.13, 12.14 and 12.15.

From Table 12.2, it is observed that efficiency of ferrite-assisted reluctance synchronous motor and reluctance synchronous motor is higher than that of induction motor for same torque output. Also, same torque is achieved in reluctance synchronous and ferrite-assisted reluctance synchronous motor with a reduction in volume in comparison with induction motor volume. Therefore, it can be concluded that ferrite-assisted reluctance synchronous and reluctance synchronous motor have better performance in terms of torque output with reduced volume and increased efficiency when compared to the induction motor.

12.5.2 Thermal Study

Thermal study involves determination of the temperature distributions in various regions of the motor. The steps involved in the coupled electromagnetic–thermal FEA are detailed in Fig. 12.16.

From Table 12.3, it is observed that the temperature of rotor is more in induction motor due to the presence of copper leading to increased losses in the rotor. Also, the reluctance motor has a thermally stable rotor as evident from the thermal analysis.

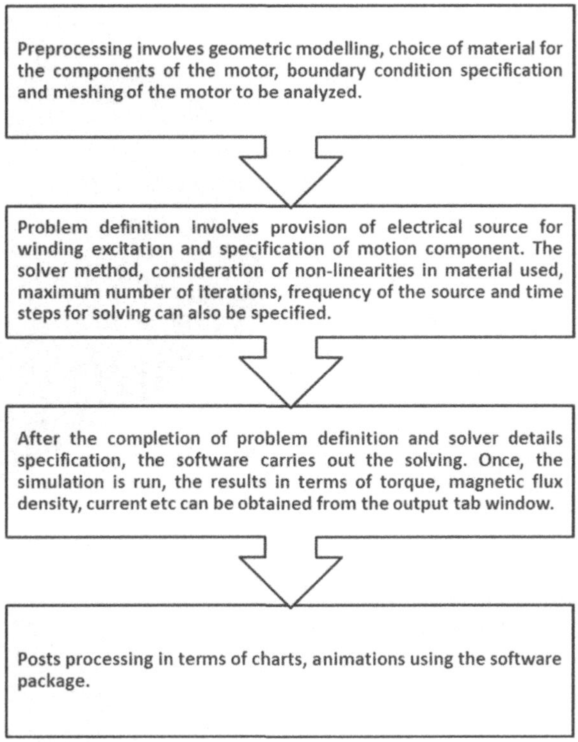

Fig. 12.12 Processes in electromagnetic FEA

Table 12.2 Electromagnetic analysis results

Performance measure/motor type	Induction	Synchronous reluctance	Ferrite-assisted synchronous reluctance
Operating speed (rpm)	1425	1500	1500
Outer diameter (mm)	254	216	216
Active stack length (mm)	254	178	140
Efficiency (%)	85	90	93
Total volume of machine (cm^3)	12,869.990	6522.357	5129.943
Torque density (Nm/m^3)	7420.36	14,641	18,616.19

Hence, with respect to thermal stability, the reluctance synchronous and ferrite-assisted reluctance synchronous motors have a thermally efficient rotor, requiring less artificial cooling compared to induction motor.

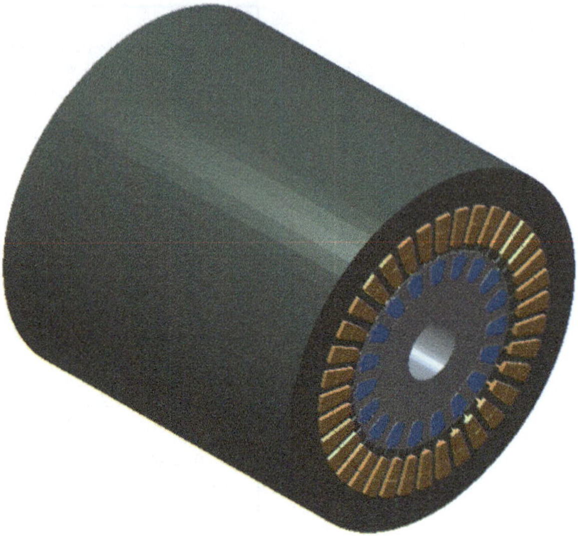

Fig. 12.13 Induction motor

Fig. 12.14 Reluctance synchronous motor

12.6 Conclusion

This paper presents an overview of centrifugal pumps along with the affinity laws, governing the operation. The operating points from the system curve with respect to speed–torque characteristics of the motor are established. A test bench was utilized for benchmarking the FEA results. Further, electromagnetic and thermal analysis was performed from which it was observed that reluctance synchronous and ferrite-assisted reluctance synchronous motors are better suited to centrifugal pump applications.

Fig. 12.15 Ferrite-assisted reluctance synchronous motor

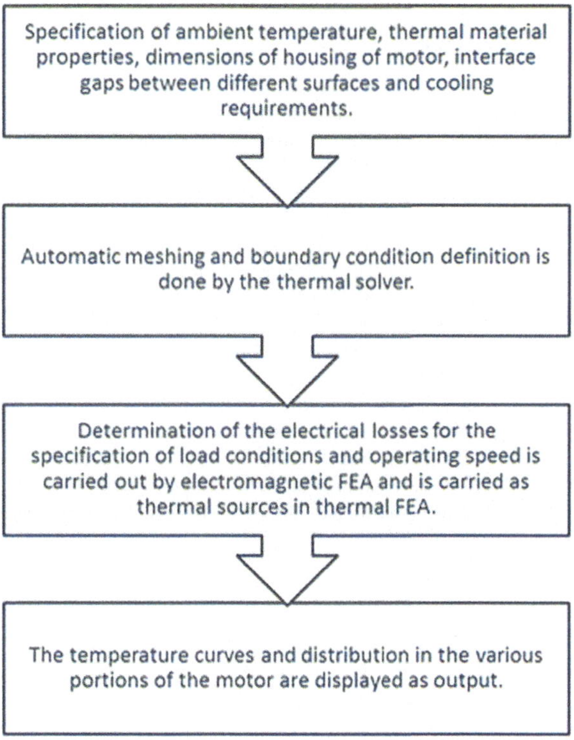

Fig. 12.16 Coupled electromagnetic–thermal FEA procedure

Table 12.3 Temperature of different parts of motor

Part/motor type	Induction	Reluctance synchronous	Ferrite-assisted reluctance synchronous
Stator winding (°C)	105	105	105
Stator core (°C)	74	73	73
Rotor core (°C)	73	61	61
Rotor copper	95 °C	Not applicable	Not applicable
Ferrite magnet	Not applicable	Not applicable	63 °C

Acknowledgements This research was supported and funded by the Department of Science and Technology, Science and Engineering Research Board, Government of India (**DST No:** SERB IRR/2016/000015) and SSN Trust.

References

1. V. Rajini, V.S. Nagarajan, *Electric Motor Drives* (Pearson, 2019)
2. W. Sölken, A Centrifugal Pump is a machine used for the purpose of transferring quantities of fluids and or gases, from one place to another. http://www.wermac.org/equipment/pumps_centrifugal.html. Last accessed 2020/07/05
3. V.S. Nagarajan, M. Balaji, V. Kamaraj, R. Arumugam, N. Ganesh, S. Srivignesh, M. Suudharshana, Design optimization of ferrite assisted synchronous reluctance motor using multi-objective differential evolution algorithm. COMPEL Int. J. Comput. Math. Electr. Electron. Eng. **36**(1), 219–239 (2017)
4. V.S. Nagarajan, V. Kamaraj, M. Balaji, R. Arumugam, N. Ganesh, R. Rahul, M. Lohit, Effect of geometrical parameters on optimal design of synchronous reluctance motor. J. Mag. **21**(4), 544–553 (2018)
5. V.S. Nagarajan, V. Kamaraj, S. Sivaramakrishnan, Geometrical sensitivity analysis based on design optimization and multiphysics analysis of PM assisted synchronous reluctance motor. Bull. Polish Acad. Sci. Tech. Sci. **67**(1), 155–163 (2019)
6. V. Rajini, V.S. Nagarajan, *Electrial Machine Design* (Pearson, 2018)
7. MagNet and ThermNet. https://www.mentor.com/products/mechanical/magnet/magnet. Last accessed 2020/07/05

Chapter 13
A Focus on Power System Congestion Management—A Primitive Review of the Literature

K. Aravindhan, M. Venmathi, and N. Chidambararaj

Abstract An increase in the involvement of distributed energy generation in the distributed systems may cause a high degree of participation of active distribution network management involving native network congestion and voltage problems. Deregulation of power system causes an increase in demand at the utility and hence results in overloading of transmission lines. To fulfill demand of power, generation companies manufacture additional output, but transmission lines are the typical ones and that they get congested as they carry additional power than its capacity. These problem areas are handled using various congestion management strategies, which play a principal role in current deregulated power systems. Hence, restructuring of power system is needed to create a shift over, where electricity is now made a artifact and has regenerated into deregulated one. In this paper, a review work is dispensed to gather related publications in congestion management. In order to reduce congestion in transmission lines, flexible AC transmission system (FACTS) devices are used, which manage the power flow to curtail losses in power system. This paper discusses the summary of congestion management in the deregulated power systems and in addition highlights the various strategies concerned to get the better of congestion in power systems.

Keywords Power system · Deregulation · Demand response · FACTS · Optimization

13.1 Introduction

In the earlier days, power systems consist of generation, transmission and distribution sectors which are arranged and defined in vertically integrated environment. There will not be any connectivity in power flow contracts of generation and distribution sectors. Due to the increase in the demand and new technological development, power

K. Aravindhan (✉)
Anna University, Chennai, India

M. Venmathi · N. Chidambararaj
St.Joseph's College of Engineering, Chennai, India

© The Author(s), under exclusive license to Springer Nature Singapore Pte Ltd. 2021

V. Kamaraj et al. (eds.), *Emerging Solutions for e-Mobility and Smart Grids*,
Springer Proceedings in Energy, https://doi.org/10.1007/978-981-16-0719-6_13

system coins the term market and hence the market has made the regulated system to deregulated environment; hence, reconstruction of industry is must with available assent present. But to apply deregulated environment there comes the overall power system that has to face lot of challenges in order to arrive the deregulated phenomenon [1]. Hence, framing of strategy toward the entire power system has to be defined in order to mitigate the issues in supplying the power to utility participants. Many problems arise to satisfy the transmission congestion, price spikes to the participants and reliability in providing the stable power flow between the different nodes and to provide stability in the markets which leads to the overall increase in the performance of the power system.

Congestion management is one of the main solutions to the versatile problems in power system network. Congestion management effectively stimulates the ideological aspects to overcome the system problems that occur during the power transfer in the deregulated market, where the interruption of many distributed renewable generators is involved [2]. The transparency is achieved in power flow Markets, and hence the power flow from suppliers to utility consumers will forego the transactions through transmission lines.

Congestion may occur due to various aspects in the power system network such as disconnection of generator, disruption of transmission line, oscillating energy demand, mismatch in transactions, in-feasibility in existing and new contracts and congestion that may also occur due to damage in system equipment. When we look on the real-time situation, changes in load demand cause many problems in overall market where it directly affects the market efficiency, and hence the consumers are demanded to turn down the power consumption, as the unit price of electricity increases. Due to congestion, the system is imposed to operate at inferior stability limits and it affects the security concern too.

The overall system might get collapse due to the occurrence of congestion, where it leads to tripping of generators in cascade and also constrains the system by tripping the system in cascade which in turn confine the system operator to further transfer power from particular generator which results in rising to charges due to congestion. Congestion has to be taken into account while solving the problems of power system to avoid wide blackouts and associated economic problems. Researches are made to overcome the drawbacks on congestion that occurs in the system, and different techniques are employed in the market to prevent the congested transmission line instead of increased electricity demand.

13.2 Overview of Congestion Management

Various methodologies are carried out in congestion management in which it is broadly divided into cost-free method and non-cost-free method [3]. The cost-free method includes disposal of transmission system operator to modify the network topology, inclusion of transformer taps and the operation of traditional systems such as phase shifter and use of FACTS devices. This method involves nominal economic consideration and does not deal with the companies related to generation and distribution sectors. The non-cost-free method deals with the rescheduling of generators and curtailment of load transactions.

In addition to two types, there are various optimization algorithms involved to make the system to operate efficiently in the markets. There are many optimization techniques such as nodal pricing scheme, price control scheme, various algorithms such as genetic algorithm, fuzzy logic, nodal and zonal congestion, particle swarm algorithm, honey bee algorithm, cuttlefish algorithm, bacterial foraging algorithm, and such many different techniques are applied to get the optimal operation of power system network. These ideas support the overall energy market to improve the extent in utilizing the asset by enhancing the available transfer capability (ATC).

The physical property of ATC is constrained over three limits of the power system; they are thermal limit, stability limit and voltage limit. Thermal limit establishes the constraint of overheating by taking into account the amount of current flowing in transmission line for a particular time period before long-lasting damage. Stability limit refers to the potential of the system to survive under perturbation during the various time periods. Due to these perturbations in overall system, the generators begin to oscillate which results in system frequency fluctuation, voltage changes, etc. The stable state can be achieved by reducing the perturbation and the fluctuations of the system to tolerable values. The range of voltage in which the system has to be maintained is indicated by the voltage limits, whereas it represents the total amount of power has been sent from one location to other.

13.3 Congestion Management Using Compensating System Devices

Congestion in transmission line can be handled by using different strategies; one such solution to congestion is done by introducing the compensating devices in transmission line. Actually, the compensating devices such as FACTS devices improve the power transfer capability of the power system by managing the transmission line impedance and by sustaining the bus magnitude and angle at both ends of the power system, thereby increasing the overall system performance and enhancing the efficiency and reliability of the system behavior on transmission line. By using FACTS devices, loading of transmission lines can be increased without violating operating constraints and hence these devices improve the available transfer capability (ATC) in power system. FACTS devices also improve the performance of the system by controlling steady-state power flow and dynamic stability without affecting the allocation of number of generators. In deregulated markets, high loading of power networks is preferred which can be done by using FACTS compensating devices. Basically, FACTS compensation is classified as series compensation, shunt compensation, combined series–shunt compensation and combined series–series compensation. Figure 13.1 shows the types of different FACTS controllers.

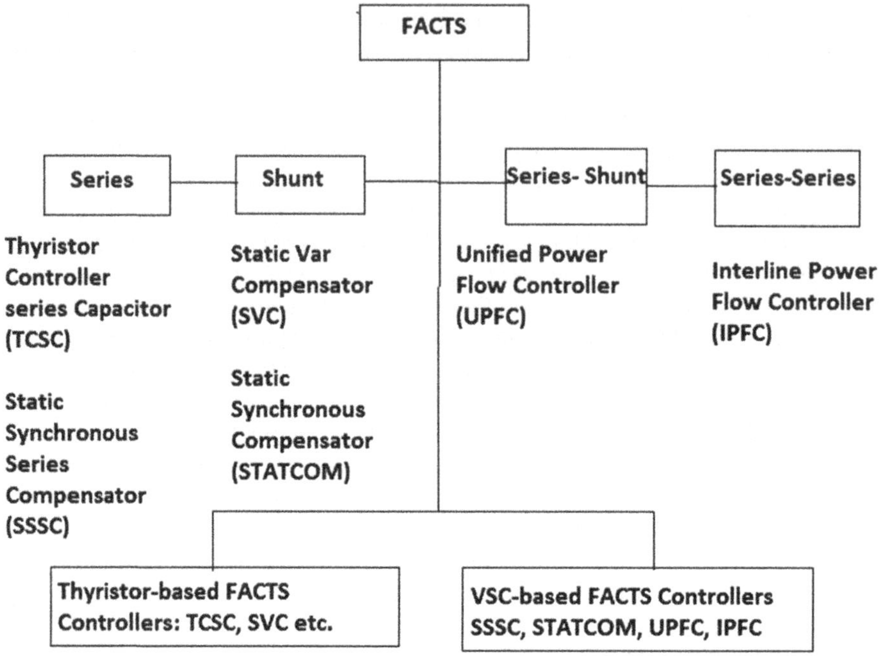

Fig. 13.1 Types of FACTS controllers

The controllers such as SVC, STATCOM, TCSC and UPFC classified above help in improving the power flow in transmission line by having the control over the reactive power and voltage in the transmission network and hence help in improving the congestion management.

The review paper proposes the use of FACTS device such as thyristor-controlled series compensator (TCSC) in the transmission line to relieve congestion [4]. The modeling of TCSC is done, and work has been carried out in IEEE 30 bus test system. Simulations are carried out, and comparative analysis of the performance of the system is calculated by using with TCSC and without TCSC. The results are tabulated and prove that placement of TCSC will surely reduce the congestion in the transmission system to the considerable extent. But in which bus the TCSC has to be placed to have effective compensation has not been discussed.

The usage of static series synchronous compensator (SSSC) in IEEE 30 bus transmission network is explained in [5], and the test case with large integration of wind farms to make congestion is also discussed. By the placement of SSSC in transmission line, the system is controlled to utilize the transmission line to maximum limit. Simulation work is carried out, but study has not explained the real-time problems in installation of SSSC in energy management system (EMS).

The overview of the literature elaborates the impact of SVC, TCSC and UPFC devices on bus voltages and power flow when the line loading is increased [6]. Comparative analysis of the three FACTS devices is done. The benefits of these

FACTS devices are illustrated by using two test systems, viz. WSCC 9 bus system and IEEE-14 bus system.

13.4 Generator Rescheduling and Load Shedding of Sensitive Buses

In the deregulated power industry, congestion is quite common and hence many techniques related to congestion management are followed based on rescheduling of generators and shedding of loads, where economic aspects are not taken into account [7]. A new term sensitivity factor (SF) is coined, and based on this factor the process of selecting highly sensitive overloaded generator and load bus are taken into account for congestion management. In the restructured grid environment, generator rescheduling or load is a primary service, where all generators and loads might not participate. The generators which are going to participate in supplying the demand and the load which are going to be utilized by the utility are prioritized in the descending order of absolute value of sensitivity factor; hence, the sensitivity factor plays a key role in selecting the generators and the loads. High sensitive buses are chosen to participate in the congestion management. Based on the necessity of utility power outputs from a generating station may be enlarged or remitting among the permissible limits, where the collaborating generator is handpicked in congestion management. On the other hand, load shedding is done in negative sensitivity buses. The sensitivity values of non-participating buses are taken zero. There is an opportunity of multiple line overloads where some buses have conflicting conditions for generation rescheduling/load shedding which are indicated for various line overloads; that is, for one overload sensitivity issue is positive and another line sensitivity factor is negative. In such cases, both the buses are chosen and improvement methodology mechanically adjusts their outputs.

The sensitivity index which relates the amendment in line current with respect to amendment in bus injections is developed to pick the collaborating generators and/or loads [8], where two conflicting objects such as tolerable overload and price of operation are decreased at the same time and congestion is administrated. But this method falls under the non-cost-free method in which inclusion of the cost is the drawback.

The frequency regulation method can be employed by applying renewable energy curtailment, by taking into account the congestion management which are within transmission network supported by price signals [9]. Since in the deregulated environment the intrusion of huge range of renewable energy sources is taking part in very important roles that cause the disturbance in frequency of the entire power system and hence creates mismatch between supply and demand. This paper deals with curtailment of the renewable energy to balance the frequency of the system which causes the ineffective use of renewable sources.

13.5 Optimization Techniques

Optimization methods are needed in the deregulated electricity markets for congestion management to achieve optimal results. Optimization methods are employed in placement of FACTS devices in the particular locality to increase the system performance in a better way. Optimization methods include different types of algorithm, one of which is mixed integer nonlinear programming (MINLP) that has been discussed with the optimal placement of series compensated device such as TCSC in transmission network [10, 11]. The MINLP was formulated to verify the situation of TCSCs within the system, and also the improvement technique was introduced to attenuate the reactive power acquisition price of the strategy. TCSC is used to get rid of congestion with minimum cost of installation by improvement approach in locating the devices. The approach uses the sensitivity analysis technique in optical location of TCSC to alleviate congestion and conjointly to enhance the facility transfer capability limit of the line.

The optimization of UPFC is introduced to minimize generation prices and relieve congestion systems. The best model for managing congestion was projected with concern for the promotion of renewable energy sources (RES) within the power grid network. It presented the wind power system with additions of FACTS devices within the system and showed improvement in transfer capabilities and congestion relief.

Soft computing techniques are the methodology in which variety of algorithm is framed to produce the effective output and to alleviate the congestion within the grid. Such technique involves the numerical algorithms using computer primarily based technologies to solve the matter associated in effective markets. Techniques involved are particle swarm optimization (PSO) genetic algorithm, multi-objective evolutionary algorithm (MOEA), cuttlefish algorithm (CFA), whale optimization algorithm (WOA), etc.

Particle swarm optimization is an improvement tool that provides a population primarily based search algorithm in which particles amend over their state with relevance time [12, 13]. It is a stochastic technique, where it deals with population primarily based search and improvement rule for finding the drawback. It is a kind of swarm principle-supported welfare function. It contributes to many engineering applications. PSO is developed based on the simulation of birds flocking in two-dimensional house. The position of each individual is going to be delineated by *XY*-axis position. Each individual is aware of the best worth up to now within the cluster among the native best. During every iteration, the answer of every particle is evaluated by the target operated. The best answer is taken as optimized solution for that iteration. PSO rule achieves the best answer for the iterative method of the algorithm. This is referred as local best fitness. The optimal placement and coordination of FACTS devices with distributed generation are made available by using particle swarm improvement technique.

The algorithm based on humpback whale known as whale optimization algorithm (WOA) rule is essentially one of the fine approaches in which especially small fishes closer to the surface area were stalked and destroyed. The bubble web feeding technique is used by humpback whales to pick its prey. In bubble web feeding algorithm, the whales emerge on a particular bubble in ring fashion. The mathematical modeling of WOA are made in three steps first finding the peripheral prey, second by applying bubble net looking technique and finally searching the prey. WOA predicts the objective prey as the prime approach since the optimum position is not recognized within the search area. After identifying the objective prey, then it follows the bubble net offensive technique which concerns shrinking peripheral mechanism and spiral position change. Next step is to obtain the new position of search agent in order to shrink and encircle the prey by finding the position between the starting and exact search agent. While moving over the spiral position, distance between the locations of whale and prey is noted down. The next movement of humpback whales in finding of prey is at random which is essentially based on their position, and in coincidence the position of search agent is streamlined in step with randomly chosen search agent rather than best search agent. WOA technique is used to resolve the position of FACTS devices within the optimum location.

In cuttlefish algorithmic rule, chromatophore cells contain red, orange, yellow, dark and earthy colored shades. Other than a lot of mirrorlike cells "iridophores and leucophores" allow cuttle skin to own all made and varied colors of its setting. The presence of the cuttle therefore depends on skin parts that have an impact on lightweight incident on the skin, light could be mirrored by either chromatophores or reflective cells or mix of each, it has the psychological changeableness of the chromatophores, and reflective cells permit the cuttle to supply a large scope of optical impacts. The calculation centers on two main methods reflection and visibility wherever reflection reproduces the lightweight reflection system, while perceivability recreates the perceivability of coordinating examples of the cuttle. These cycles are utilized in search strategy to look for out the global optimum answer. Cuttlefish algorithmic rule technique is utilized in looking of optimum bus and additionally in choice of generators in deregulated business sectors or markets.

Cuckoo search (CS) is an improvement algorithmic program impressed by the brood interdependence of cuckoo species, which lay their eggs in the homes of alternative host feathered creatures. It has been applied into the designing optimization issues and demonstrated its promising potency. In the host species home's cuckoo winged the animal to lays her eggs. In the presumption of the eggs are its own by the host, it deals with it. While the host discovered that an egg is not its own, it might destroy either the egg or the nest so construct a replacement nest at a special location. In the recent arrangement by changing the characteristics, new arrangements are generated in flip to increase the standard of arrangements. A cuckoo egg speaks to another solution, whereas every host fowl egg in a very nest represents an answer. The point is to supplant a most exceedingly terrible solution within the nests with the new and probably higher solutions. Like this, there are several improvement techniques that illustrate to solve the important time drawback of congestion within the deregulated power markets.

The non-cost free method of rescheduling generation (or re-dispatch optimization) issues is solved by exploiting a change over of particle swarm optimisation (PSO) rule [14]. Selection of the participation generators is involved to relieve congestion.

Based on the PSO, the selection of particular group of generators is operated to make the system more reliable and to have effective system operation. But choosing the best outcomes varies with the initialization of variables.

The Cuckoo Search method in comparison with the other existing ways like fuzzy adaptative bacterial foraging (FABF), simple bacterial foraging (SBF), particle swarm optimisation (PSO) are done wherever optimization of the system is underneath required for additional governance over the comparative study [15].

The Optimal power flow primarily based congestion management (OPF based CM) overcome the drawbacks to minimize the value and by calculating Locational Marginal Pricing (LMP) and finally relieve network congestion in the systems. Here the different costs are chosen supported the consumption of power by the utility is finished in how that costs are enlarged throughout peak masses that ultimately cause the utility to be on top of things over the consumption of power that finally relieve the facility congestion within the line [16].

13.6 Conclusion

The review based on congestion management in the deregulated power market is proposed in this paper. Initially, it starts with defining the congestion management and ends up with different optimization techniques involved in sorting out the problems to avoid congestion. Still now, many real-time problems are faced by the transcos and discos in the restructured power market to provide effective power transfer to reduce losses. The paper also describes the utilization of FACTS devices to reduce the congestion and to increase the effective power transfer in the congested bus. Hence, the solution to the difficulties faced by congestion is surrendered to the researchers to extend their work and to come up with smarter solutions. The effective use of the asset is encouraged in solving the problems associated with the congestion management. The problem can also be solved by choosing the variables in network or in the load side.

References

1. F. Flores-Espino, T. Tian, I. Chernyakhovskiy, M. Mercer, *Competitive Electricity Market Regulation in the United States: A Primer*. National Renewable Energy Lab., Denver, CO, Tech. Rep. NREL/TP-6A20-67106 (2016)
2. Z. Zhang, Y. Ren, M. Zhou, A comprehensive decision model of optimal Demand Response program, in *5th International Conference on Electric Utility Deregulation and Restructuring and Power Technologies* (Changsha, China, 2015), pp. 121–126
3. A. Pillay, S. Prabhakar Karthikeyan, D.P. Kothari, Congestion management in power systems—a review. Int. J. Electr. Power Ener. Syst. **70**, 83–90 (2015)
4. P. Choudekar, S. Sinha, A. Siddiqui, Congestion management of IEEE 30 bus system using thyristor controlled series controlled series compensator, in *International Conference on Power Energy, India* (2018)

5. L. Yao, P. Cartwright, L. Schmitt, X.-P. Zhang, Congestion management of transmission systems using FACTS, in *IEEE/PES Transmission and Distribution, International Conference, China* (2005)
6. A.S. Siddiqui, T. Deb, Congestion management using FACTS devices. Int. J. Syst. Assur. Eng. Manag. (2013)
7. N. Chidambararaj, K. Chitra, A combined congestion management technique for rescheduling of optimal active power of generator units. J. Theor. Appl. Inform. Technol. **67**(3) (2014)
8. J. Hazra, A.K. Sinha, Y. Phulpin, Congetion management using generation resheduling and/or load shedding of sensitive buses, in *3rd Internal Conference on power Systems, India* (2009)
9. H. Bae, T. Tsuji, T. Oyama, Frequency regulation method with congestion management using renewable energy curtailment, in *IEEE Conference* (2006)
10. P.P. Kulkarni, N.D. Ghawghawe, Optimal placement and parameter setting of TCSC in power transmission system to increase the power transfer capability, in *International Conference on Energy Systems and Applications (ICESA 2015), India* (2015), pp. 735–739
11. L.N. Mrunalini Devi1, A. Surya PrakashRao, Optimal placement of TCSC for reactive power reserve management with reactive power loss minimization using hybrid Psogsa. Int. J. Power Syst. Microelectron. (TJPRC: IJPSM) **1**(1), 61–72 (2016)
12. H. Mahala, Y. Kumar, Active & reactive power rescheduling for congestion management using new PSO stretagy, in *IEEE Students' Conference on Electrical, Electronics and Computer Science (SCEECS)* (2016), pp. 1–4
13. S. Dutta, S.P. Singh, Optimal rescheduling of generators for congestion management based on particle swarm optimization. IEEE Trans. Power Syst. **23**(4), 1560–1569 (2008)
14. S. Abinaya, N. Chidambararaj, K. Aravindhan, Optimal placement and co-ordination of UPFC with DG using PSO, J. Adv. Res. Dyn. Control Syst. **10**(15) (2018)
15. N. Chidambararaj, K. Chitra, Minimal constraint based Cuckoo search algorithm for removing transmission congestion and rescheduling the generator units. Int. J. Bus. Intell. Data Mining **14**(1/2) (2019)
16. R. Guguloth, T.K. Sunil Kumar, LMP calculation and OPF based congestion management in deregulated power systems, IEEE (2016)

Chapter 14
Battery Safety Enhancement in Electric Vehicles—A Review Paper

R. J. Vijaya Saraswathi and V. Vasan Prabhu

Abstract In the present scenario, electric vehicles and hybrid electric vehicles are fast replacing the combustion engine-type vehicles. Due to this sudden surge in growth of electric vehicles, there arises a major concern for the continuous overseeing and the need for safety enhancement features for battery, drivetrain and what should be done in case of emergencies. There are numerous ongoing works and efforts in the area of electric vehicles in order to make them cost effective and increase their efficiency. One of the main factors to be considered for the failure-free operation of vehicles is the batteries used. In this comprehensive literary review, the working and safety aspects of lithium-based battery used in electric vehicles are analysed. Safety goals for the safe operation of electric vehicles with respect to the batteries and battery management systems (BMS) involved are discussed. Various methods and technologies adopted and proposed to achieve these goals are analysed here.

Keywords Electric vehicles · Lithium-based battery · Battery management system

14.1 Introduction

Due to alarming levels of increase in pollution and fast depletion of fossil fuels, restriction of vehicle emissions and high demand for fossil fuels, automobile manufacturers around the globe are looking into new car model that would work in an alternative way efficiently. Thus, taking into view the rate of depletion of fossil fuels, electric vehicle (EV) has been developed to make our planet a better place to live. The function of a battery in electric vehicle or hybrid electric vehicle plays a major part in the operation and performance of electric vehicles. This paper focuses on

R. J. Vijaya Saraswathi (✉)
Panimalar Engineering College, Chennai, India

V. Vasan Prabhu
Department of Automobile Engineering, SRM Institute of Science and Technology (Deemed to be university), Kattankulathur, Tamilnadu, India
e-mail: vasanprv@srmist.edu.in

© The Author(s), under exclusive license to Springer Nature Singapore Pte Ltd. 2021
V. Kamaraj et al. (eds.), *Emerging Solutions for e-Mobility and Smart Grids*,
Springer Proceedings in Energy, https://doi.org/10.1007/978-981-16-0719-6_14

Table 14.1 Comparison between various lithium-ion technologies (on a scale of 4)

	Expenditure	Life span	Performance	Safety	Specific power	Specific energy
Lithium–nickel–cobalt aluminium	2	4	3	2	4	4
Lithium–nickel–manganese–cobalt	3	3	3	3	3	4
Lithium–manganese spinel	3	2	2	3	3	3
Lithium–titanate	1	4	4	4	3	2
Lithium–iron phosphate	3	4	3	4	3	2

various techniques involved in making a battery a reliable source, recognizing the failure types, identifying the levels of safety required and the methods to achieve it.

Considering the advantages of lithium-ion (Li-ion) batteries like very less self-discharging rate of only 5% per month, higher value of specific energy (140 Wh/kg) and higher value of energy density, Li-ion batteries have emerged as a standard for today's battery electric vehicles [1]. Table 14.1 shows a comparison between various lithium-ion technologies.

14.2 The Current Scenario of Batteries in Electric Vehicles

All electric vehicles have power electronics components, electric motor, battery, port, connector and charging setup as shown in Fig. 14.1 [2]. There are a lot of ongoing researches in making each one of these components more efficient [3]. The battery plays a major role in traction system of electric vehicles. Electric vehicle

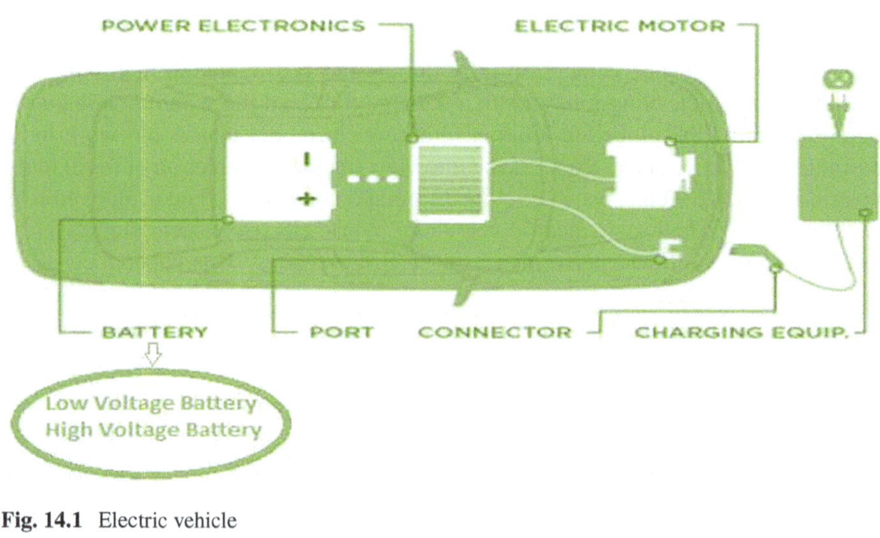

Fig. 14.1 Electric vehicle

charging mechanism is of two types. In the first type, the charger is present internally which works on the standard charging process. And the other type is a fast charger used externally. The fast charger provides a very quick charging time but reduces its life time. The low-voltage battery is used for powering up of low voltage units, control units, steering systems, etc. The high-voltage battery is used for powering up the electric traction motor. Every battery used in electric vehicles undergoes the following seven steps in a cycle. They are: manufacturing of components, bringing the components together in the form of cells, grouping the cells into modules, assembling and packaging, integrating it to electric vehicles, using it in its life time, recycle and reuse [4].

The cells of the battery are welded together in series or parallel pattern or a combination of both to acquire the required power capability. Batteries are arranged in an array to overcome the space constraint and also for the integration of thermal management. These arrays of batteries are arranged inside a metallic battery case, which should also hold the thermal management system and battery management system (BMS) [5]. A few examples of batteries used in electric vehicles are shown in Fig. 14.2.

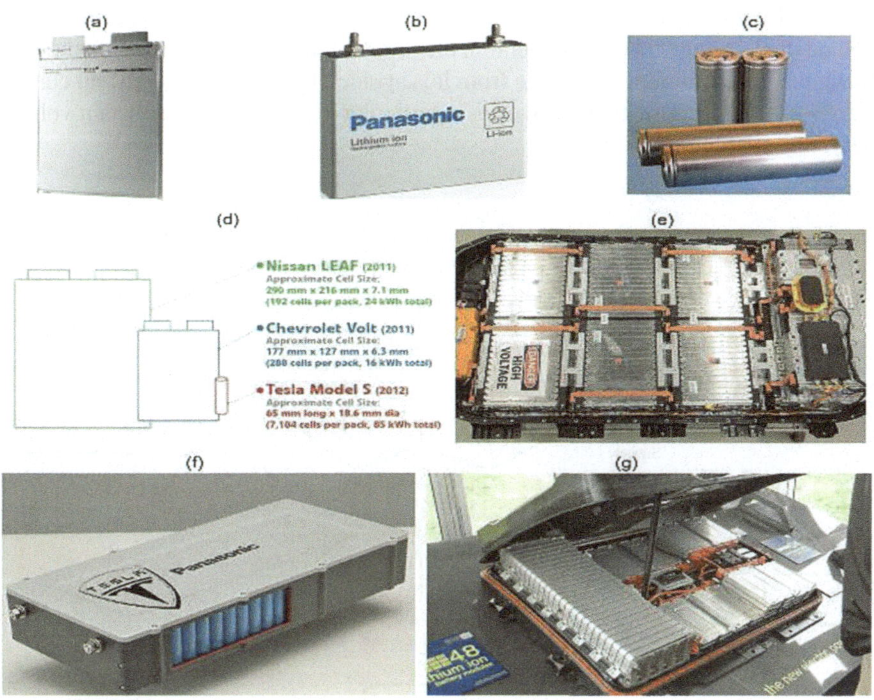

Fig. 14.2 Batteries used in electric vehicles: **a** pouch battery, **b** prismatic battery, **c** the cylindrical model, **d** comparison of size of types of batteries, **e** 288 prismatic type batteries combined to give 60 kWh by LG Chem, **f** number of cylindrical batteries combined to form a battery pack by Tesla and **g** leaf-type battery (Tesla)

14.3 Battery Management System

The function of BMS is to monitor the voltage and current of each cell, state of charge (SOC), provide on–off signal for safety devices, provide battery status and diagnostic data to the PC and provide on–off signals for heating and cooling elements. The functions of BMS are shown in Fig. 14.3.

The various methods found in the literature for battery charging are constant current (CC) method, constant voltage (CV) method and constant current–constant voltage (CCCV) method. The current is maintained at same value in all the series-connected batteries in the CC method. When the SOC raises, the internal resistance of the batteries also tends to raise. So to maintain the charging at a constant rate, the voltage must be increased [6]. One has to be careful in the selection of charging current because, if the charging value of current is very high then it may result in the battery getting charged quickly, but it may also lead to overcharging and overheating. In the second method of charging, i.e. CV type, a larger amount of current is allowed to be drawn as the battery has very low initial resistance. After reaching float voltage, i.e. a safe level of voltage, there is a gradual decrease in current value. As the SOC of the battery starts increasing, the resistance of the battery also starts reaching higher values, which leads to decrease in the current drawn by the battery [7]. In the last method of charging, i.e. CCCV method, the problem of overcharging the batteries or drawing a larger amount of voltage from it is avoided by employing constant current charging method till the battery reaches a particular level. Then, the constant voltage method is continued [8].

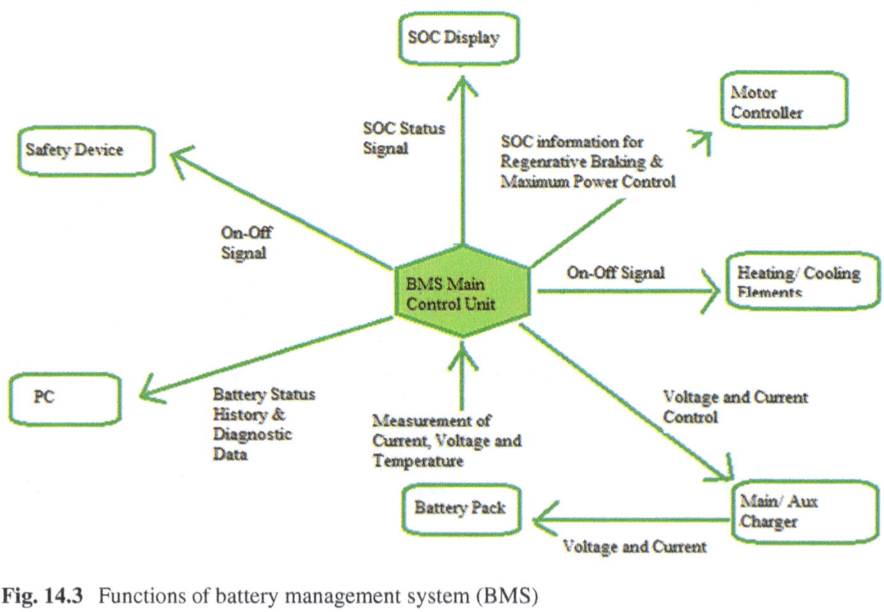

Fig. 14.3 Functions of battery management system (BMS)

Table 14.2 Types of charging

	Level-1 type (AC)	Level-2 type (AC)	Fast charging (DC)
Volts (V)	1-phase AC, 120 V	1-phase AC, 208/240 V	208/480 V 3-phase AC
Amperes (A)	12–16	12–80	<200
Charging loads (kW)	1.4–1.9	2.5–19.2	<150
Time taken for charging	3–5 miles of range/hour	10–20 miles of range/hour	80% charge in <30 min

14.4 Types of Charging

Charging can be categorized as follows [9] (Table 14.2).

14.5 Improving the Safety Aspects of Batteries

14.5.1 BMS and Cell Balancing

An effective battery management system should ensure safe operation of the vehicle, assure a long time reliability and prevent damage of the battery and keeping the battery cost effective.

When excess amount of current is drawn by a battery, its temperature increases which also affects the output voltage. BMS has to perform the function of limiting the current in the constant current model [10].

The battery management system has to maintain the balance in the level of voltage in each of the cells in the pack to prevent overcharging and improve the efficiency of the pack. BMS should be capable of passing on the charge from one cell to another in case if they are charged unevenly, thus providing cell balancing. The charge in the cells can be maintained evenly by passive type of balancing or active type of balancing. The passive type dissipates charge from the cells having excess charge in the form of heat. The active method involves transferring charge from the cells that have high charge to the ones that have low charge. The active method involves the usage of either capacitor or transformer or inductors or converters for cell balancing.

In the capacitor-based method, cell balancing is achieved by using a capacitor. From the cells that have high voltage, the charges are transferred to the ones that have low voltage. A capacitor is connected through switches which connect it to high-voltage cell first, and then the switch connects it to low voltage cell. In the inductor-based method, the charge from the cell that has a larger amount of voltage is pushed into the inductor and then discharged to the cell that has a lesser amount. It is faster and more efficient than the capacitor method. In the transformer method, rapid switching of the primary coil takes place and diodes are positioned to control the current flow. This type of balancing is effective than passive balancing and faster than capacitive methods since balancing takes place automatically and does not involve the use of any complex procedures [11].

In the converter-based system, a DC-to-DC converter is employed to transfer energy between the higher and lower energy cells [12]. It can be broadly classified into three depending on the type of transfer taking place, from the pack to a cell, from the cell to a pack and from cell to cell. In the pack to cell category, energy is transferred from the entire pack to a single cell that has less charge. In the cell to pack category, the energy is transferred from a single cell that has more energy to the pack of other cells. In the cell to cell category, transfer of energy takes place from one particular cell to another cell.

14.5.2 Analysing the Hazards and Risks Involved

According to ISO26262 standard, hazard and risk assessment is to be carefully analysed. So, the major hazards and risks of the usage of battery packs are to be analysed. Fire, shock and sudden shortage of battery when the vehicle is operated are the major risks involved with the batteries. The major risks can be classified under the following criterions, namely exposure (Exp), severity (Sev), controllability (Con). Exp1 (incredible) to Exp4 (very probable) are the levels of exposure. Sev1 (not wounded) to Sev3 (disastrous) are the levels of severity. Con1 (manageable/controllable) to Con3 (hard to manage/control) are the levels of controllability.

These criteria, according to various combinations between them, are put together in the form of matrix, with an automotive safety integrity level (ASIL) as given in the risk graph matrix of ISO26262, Table 14.3. Quality management or Q is the first level in ASIL that requires no special safety integrity requirement. The other levels are A, B, C and D, with type A being the least limiting type and type D being the very constrictive type ASIL.

Application-specific hazard and risk assessment (HaRA) has to be done, and its results have to be categorized. From these results, safety goals for that particular application are to be determined. Then, an architecture to achieve these safety goals is developed [13].

14.5.3 Prognostics and Health Management (PHM) of Batteries

PHM is widely used to predict the fault conditions that may arise in future. Prediction of RUL of batteries and analysing the SOH of batteries plays a major role in user safety and crucial decision-making. In [14], Choi et al. proposed a framework

Table 14.3 Risk graph matrix of ISO26262

Severity	Exposure	Controllability		
		Con1	Con2	Con3
Sev1	Exp1	Q	Q	Q
	Exp2	Q	Q	Q
	Exp3	Q	Q	A
	Exp4	Q	A	B
Sev2	Exp1	Q	Q	Q
	Exp2	Q	Q	A
	Exp3	Q	A	B
	Exp4	A	B	C
Sev3	Exp1	Q	Q	A
	Exp2	Q	A	B
	Exp3	A	B	C
	Exp4	B	C	D

for estimating the capacity of lithium-ion battery using the multi-channel data of voltage (V), current (I) and temperature (T). The machine learning techniques such as feedforward neural network (FNN), convolutional neural network (CNN), long short-term memory (LSTM) networks. Data sets of NASA for lithium-ion battery were taken for analysis, and it was found that LSTM showed the best performance, followed by FNN and CNN. This multi-channel method has shown an improved performance over single data voltage of 58, 46 and 25% mean absolute percentage error with respect to FNN, CNN and LSTM.

14.5.4 Feature Vector Method

If the SOH of a battery falls below 80%, then it is considered unsafe for usage. So, the important feature of any BMS has to be online estimation of SOH to provide safe operation of electric vehicles [15, 16]. In [17], Naha et al. proposed collection of partial charging data. From this data, a feature vector comprises the difference in the values of voltages obtained and average temperature. This feature vector is developed for varying levels from 100 to 80% of SOH. Various methods are tried, out of which artificial neural network (ANN) model gives the best link connecting the desired SOH and the obtained feature vectors. This method collects its data from the already present BMS. Four hundred sets of data obtained during the charging cycle and discharging cycle are required for training this algorithm. And for the estimation of SOH, it approximately takes 15 min of usage data. Its error is found to be less than 1% mean absolute error. In [18], an analysis on the combination of support vector regression technique and feature vector technique is done and the results show that the number of data required for training is decreased and predicting the RUL efficiency has increased.

14.6 Future Scope and Conclusion

Batteries used in EVs require a major care for a failure-free operation. The BMS of an EV monitors and takes care of all the vital parameters related to the batteries. Major recent technologies in improving the safety aspects of batteries are analysed here in this paper. There are many technologies like electrochemical model-based methods, equivalent electrical circuits method and data-driven methods to estimate the health of the batteries. Based on various literatures, it is found that data-driven methods like deep learning and machine learning are found to be very effective in estimation and maintain the battery health, as these methods do not need in-depth battery details. It is found that training of the algorithms or models for these data-driven methods are also easier and the error rate is also comparatively lesser.

References

1. Batteries for Electric Cars Challenges, Opportunities and the Outlook to 2020 by the Boston Consulting Group
2. https://www.driveelectricvt.com/charging-stations/types-of-charging
3. Ms. Anbu, M. Sukanya, Mr. V. Vasan prabhu, FLC based controller for a dual mode operation of 8/6 SRM with operation at optimum angle for enhanced operation in hybrid electric vehicles. Int. J. Appl. Eng. Res. **10**, 16344–16348 (2015)
4. P. Sun, R. Bisschop, Huichang, X. Huang, A review of battery in electric vehicles, in *A Review of Battery Fires in Electric Vehicles, Fire Technology*, vol. 56 (2020)
5. https://batteryuniversity.com/index.php/learn/article/types_of_battery_cells
6. C. Liu, K.T. Chau, D. Wu, S. Gao, Opportunities and challenges of vehicle-to-home, vehicle-to-vehicle, and vehicle-to-grid technologies. Proc. IEEE **101**, 2409–2419 (2013)
7. F. Ahmad, M.S. Alam, M. Asaad, Developments in EVs charging infrastructure and energy management system for smart microgrids including EVs. Sustain. Cities Soc. **35**, 552–564 (2017)
8. F. Mwasilu, J.J. Justo, F.K. Kim, T.D. Do, J.W. Jung, Electric vehicles and smart grid interaction: a review on vehicle to grid and renewable energy sources integration. Renew. Sustain. Energy Rev. **34**, 501–516 (2014)
9. M. Parchomiuk, A. Moradewicz, H. Gawiński, An overview of electric vehicles fast charging infrastructure, 2019 Progress in *Applied Electrical Engineering (PAEE)*, Koscielisko, Poland, 2019, pp. 1–5. https://doi.org/10.1109/PAEE.2019.8788983
10. S. Li, C. Zhang, S. Xie, Research on fast charge method for lead-acid electric vehicle batteries, in *Proceedings of International Workshop on Intelligent Systems and Applications*, Wuhan, China, 23–24 May 2009, pp. 1–5
11. M.M. UrRehman, M. Evzelman, K. Hathaway, R. Zane, G.L. Plett, K. Smith, E. Wood, D. Maksimovic, in *Modular Approach for Continuous Cell-level Balancing to Improve Performance of Large Battery Packs*, ECCE2014
12. F. Baronti, M. Chow, C. Ma, et al., E-transportation: the role of embedded systems in electric energy transfer from grid to vehicle. J. Embed. Syst. 10 May 2016 [Springer Open]

13. D. Marcos, J. Perez, P. Zubizarreta, M. Garmendia, I.P. de Arenaza, J. Crego, J.A. Cortajarena, A safety concept for an automotive lithium-based battery management system, in *EV2019, Electric Vehicles International Conference & Show* (2019)
14. Y. Choi, S. Ryu, K. Park, H. Kim, Machine learning-based lithium-ion battery capacity estimation exploiting multi-channel charging profiles, special section on advanced energy storage technologies and their applications, in *IEEE Access*, vol. 7 (2019)
15. D. Zhou, H. Yin, W. Xie, P. Fu, W. Lu, Research on online capacity estimation of power battery based on ekf-gpr model. J. Chem. (2019)
16. X. Li, Z. Wang, State of health estimation for lithium-ion battery by combing incremental capacity analysis with gaussian process regression (2019)
17. A. Naha, S. Han, S. Agarwal et al., An incremental voltage difference based technique for online state of health estimation of Li-ion batteries. Sci. Rep. **10**, 9526 (2020)
18. Q. Zhao, X. Qin, H. Zhao, W. Feng, A novel prediction method based on the support vector regression for the remaining useful life of lithium-ion batteries, Microelectron. Reliab. **85**, 99–108 (2018). ISSN 0026-2714

Chapter 15
Performance Evaluation of Fast Charging DC–DC Converter for Electrical Vehicle with Root Locus-Based Digital Controller

Balakrishna Nallamothu, B. Santhana Krishnan, and Ravindra Janga

Abstract In recent times, because of range anxiety, charging time and charging infrastructure of electric vehicles (EVs) are the main subjects to wide gap in research and development. Charging station infrastructure plays a major role in easy transition from traditional IC engine-based vehicles to EVs. The implementation of off-board fast charging for EVs with proposed control strategy aimed to facilitate the improved performance of a DC–DC converter. However, there are many analog controllers proposed in controlling converters, which lack in optimum operation of high-frequency DC–DC converter. Digital controllers provide optimum control not only in case of static input voltage but also in case of change in input voltage. At first step, pole–zero mapping is used in designing the small signal-based discrete-time model of the desired DC–DC converter. Based on discrete-time modeling, digital controller is designed using root locus technique. Lastly, simulation results are acquired for a 400 V/120 A DC–DC converter, which is used in fast charging system of electric vehicle. The analysis of the proposed controller design is presented in detail.

Keywords Electric vehicle · Fast charging · DC–DC converter · Pole–zero mapping · Root locus technique

15.1 Introduction

Electric vehicles (EVs) have grown exponentially due to its high efficiency and eco-friendly benefits. The increase in the sales of electric vehicles increases the technical requirements for fast charging. Due to constraint of the interior space, charging necessities should meet high charging efficiency. The strongest recommendation to overcome these barriers is to install off-board charging facilities primarily known as charging stations. These charging stations are fed from multiple

B. Nallamothu (✉) · B. Santhana Krishnan
Electrical Department, Annamalai University, Annamalai Nagar, Chidambaram, India

R. Janga
Department of EEE, Bapatla Engineering College, Bapatla, India

© The Author(s), under exclusive license to Springer Nature Singapore Pte Ltd. 2021

V. Kamaraj et al. (eds.), *Emerging Solutions for e-Mobility and Smart Grids*,
Springer Proceedings in Energy, https://doi.org/10.1007/978-981-16-0719-6_15

sources of energies from either solar, wind or fuel cell. Charging stations supplied from renewable energy sources further decrease the carbon levels in the environment compared to stations integrated to conventional grids. However, the output voltage from these renewable sources is variable in nature, which leads to the importance for controlling DC–DC controller effectively to maintain the output voltage constantly with change in input by the source. There are some trade-offs in designing high switching frequency converters between faster transient response and efficiency. Selection of proper controllers plays an important role in achieving highest efficiency from the DC–DC converter. Digital controller has positive edge over analog controller in controlling high frequency converters. The advantage of digital controller that comes with programming is seen missing with the analog controllers. Recent trends in the digital controllers show the increase in the efficiency of the various converters. The digital controllers also offer positive edge in controlling high-frequency switching converters in various applications of automobile, communications and signal processing.

This leads to the focus on the design of efficient digital controllers, which work on the basic power electronic component DC–DC converter of an electric vehicle charging station. Different systems have been designed to implement charging station models, which fed from solar power and other renewable sources. Some progress has been seen from [1–5], toward implementing the design of charging station models using DC–DC converters, which are fed from different renewable sources. One of the efficient topologies proposed in [6] intended for high input voltage and below medium power needs. DC–DC converter model proposed regulates the output voltage for variable loads. A novel approach has been proposed in [6] using constant current controller. In [7, 8], design of digital controller based on root locus technique has been proposed to control in more efficient manner. The concept of small signal modeling is proposed in deriving mathematical model in [9, 10, 11]. But, no one has proposed a design to implement a digital controller for controlling a DC–DC converter used in EV charging application.

The main idea of the paper is to determine the discrete-time domain model of the buck converter by implementing a design procedure using a root locus-based digital controller. In Sect. 15.2, overall system description is given along the working model of the DC–DC converter. In Sect. 15.3, design procedure of discrete-time domain model of the resonant converter is resolved using pole–zero mapping technique and analysis of the digital controller is formulated. Lastly, confirmation of the results obtained by theoretical and modular analysis using MATLAB of the proposed root locus-based controller is presented in Sect. 15.4 and conclusive remarks of the designed system are given in Sect. 15.5.

15.2 Working Principle of the Considered System

The schematic diagram shown in Figs. 15.1 and 15.2 illustrates the feeding of different types of renewable energy sources to a system, which comprises DC–DC

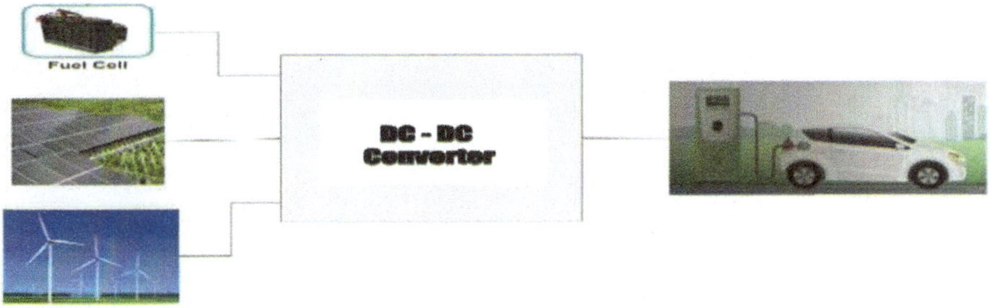

Fig. 15.1 Overview of the considered system

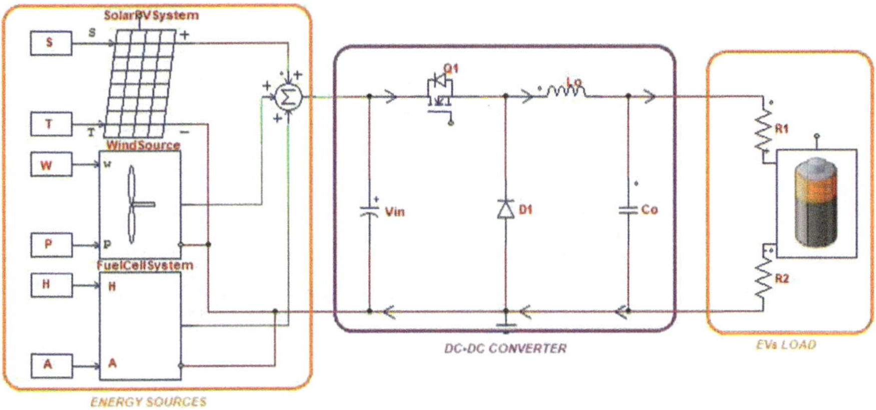

Fig. 15.2 Internal overview of the considered system

converter. The output voltage of the combined energy sources is derived as 800 V with a rating of 48 KW, which is shown in Fig. 15.3. The main working principle of DC–DC converter is to handle variable voltage at the input side and generate fixed voltage to the load side. Along with the usage of renewable energy sources, performance of the converter also has equal importance.

In present work, the output current rating of the DC–DC converter is raised to above 100 A to facilitate fast charging of electric vehicle. Figure 15.4 depicts an illustration of a basic buck converter and ideal case VI characteristic waveforms for the inductor voltage and current. The transistor (Q1) is considered operating at a constant frequency 'fs' and an assumed duty ratio of 'd'. The other assumptions considered are continuous inductor current, lossless circuit components at all conditions and infinitesimal ripple in output capacitor voltage. The relationship between operating output voltage and duty ratio is derived by formulating the positive and negative inductor volt second in one complete switching cycle.

In steady-state operation, equation for the input voltage is given as:

$$(V_{\text{in}} - V)dT_s = V(1 - d)T \Rightarrow \text{V} = \text{d}V_{\text{in}} \tag{15.1}$$

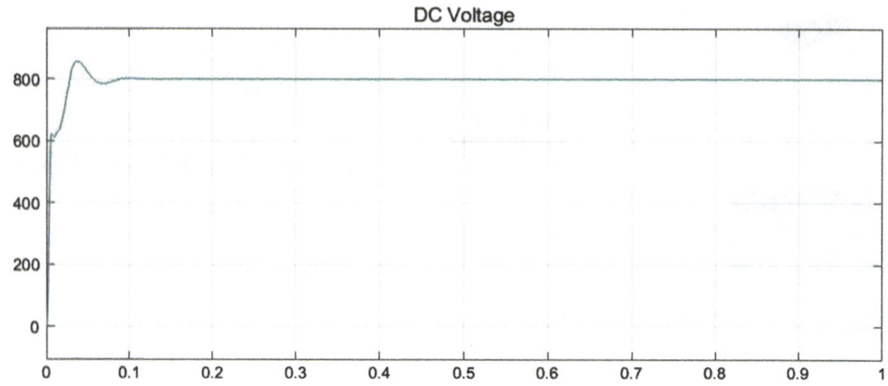

Fig. 15.3 Output voltage wave form of combined energy sources

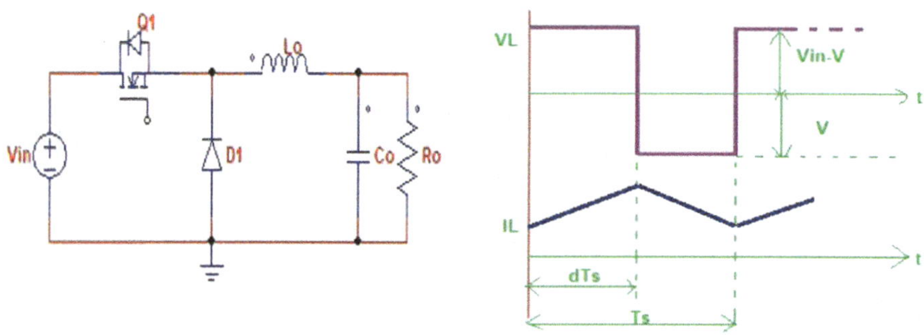

Fig. 15.4 DC–DC converter with output waveforms

where $d = T_{\text{on}}/T_s$.

15.3 Mathematical Model of the Converter

The necessity of controller for the DC–DC converter is to withstand the output voltage constant value irrespective of any considerable changes in the DC source input voltage Vin and the output battery load current. So firstly, in designing a digital controller for any type of converter it starts by solving mathematical equations which gives relation between input–output and parameters which control the converter.

Out of all other techniques, state space averaging offers additional edge and facilitates here to mathematically model the DC–DC converter. In this chosen method, state space equations are derived for each of the two modes, i.e., during ON state and OFF state in one complete operation cycle. Two modes can be derived when the considered converter operating in the continuous current mode, thereby producing two separate space equations when the converter operating in two states. Figure 15.5 shows the circuit diagram of the DC–DC converter corresponding to one of the states, i.e., ON state of power switch, and the corresponding state space equations for the ON state are specified as:

$$\begin{bmatrix} \dot{i} \\ \dot{v} \end{bmatrix} = \begin{bmatrix} 0 & \frac{-1}{L} \\ \frac{1}{C} & \frac{-1}{RC} \end{bmatrix} \begin{bmatrix} i \\ v \end{bmatrix} + \begin{bmatrix} \frac{1}{L} \\ 0 \end{bmatrix} [v_{\text{in}}] \tag{15.2}$$

Figure 15.6 illustrates the circuit arrangement of the buck converter related to OFF state of power electronic switch, and respective sate space equations are

$$\begin{bmatrix} \dot{i} \\ \dot{v} \end{bmatrix} = \begin{bmatrix} 0 & \frac{-1}{L} \\ \frac{1}{C} & \frac{-1}{RC} \end{bmatrix} \begin{bmatrix} i \\ v \end{bmatrix} + \begin{bmatrix} 0 \\ \frac{1}{C} \end{bmatrix} [v_{\text{in}}] \tag{15.3}$$

The output voltage of the converter 'v' and the inductor current 'i' are the elements of a state vector x, whereas the input vector u has V_{in}. By considering a weighted average of the equations of Figs. 15.5 and 15.6, state space averaged model of the converter is formulated as

$$\dot{x} = Ax + Bu \tag{15.4}$$

where

$$A = dA_{\text{ON}} + (1 - d)A_{\text{OFF}}$$
$$B = dB_{\text{ON}} + (1 - d)B_{\text{OFF}}$$

Therefore, the averaged matrices for the buck converter are

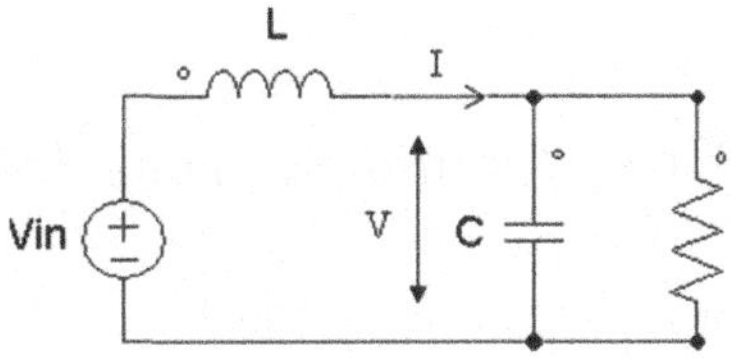

Fig. 15.5 Equivalent to the converter during ON state of switch

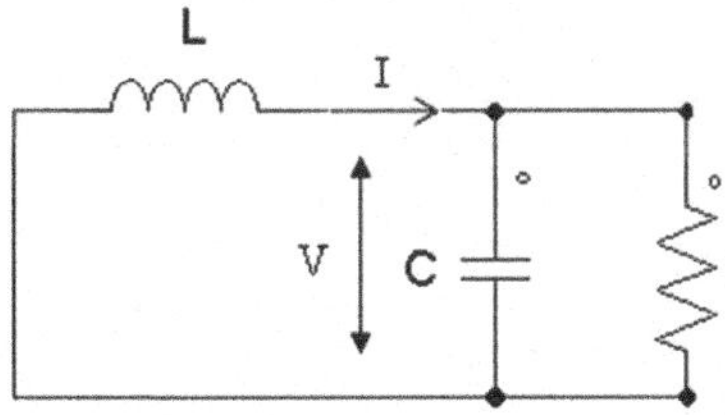

Fig. 15.6 Equivalent to the converter during OFF state of switch

$$A = \begin{bmatrix} 0 & \frac{-1}{L} \\ \frac{1}{C} & \frac{-1}{RC} \end{bmatrix} \quad B = \begin{bmatrix} \frac{d}{L} \\ 0 \end{bmatrix} \tag{15.5}$$

The duty ratio being the control input to the converter is in matrix *B*. Therefore, the averaged model is time varying and difficult to solve. To simplify the proposed model, Eq. 15.4 is linearized, by considering small variations in the variables.

Each variable is sum of DC component and small signal component as

$$\begin{matrix} x = X + \hat{x} & & \hat{x} \prec\prec X \\ u = U + \hat{u} & \text{where} & \hat{u} \prec\prec U \\ d = D + \hat{d} & & \hat{d} \prec\prec D \end{matrix} \tag{15.6}$$

By neglecting higher order terms and on substituting above values in Eq. (15.4), we get,

$$\dot{\hat{x}} = A\hat{x} + B\hat{u} + (A_{\text{on}} - A_{\text{off}})\hat{d}X + (B_{\text{on}} - B_{\text{off}})\hat{d}U \tag{15.7}$$

$$\dot{\hat{x}} = A\hat{x} + B\hat{u} + E\hat{d} \tag{15.8}$$

where $E = (A_{\text{on}} - A_{\text{off}})X + (B_{\text{on}} - B_{\text{off}})U$.

And the state model of buck converter is

$$\begin{bmatrix} \dot{\hat{i}} \\ \dot{\hat{v}} \end{bmatrix} = \begin{bmatrix} 0 & \frac{-1}{L} \\ \frac{1}{C} & \frac{-1}{RC} \end{bmatrix} \begin{bmatrix} \hat{i} \\ \hat{v} \end{bmatrix} + \begin{bmatrix} \frac{1}{L} & \frac{V_{\text{in}}}{L} \\ 0 & 0 \end{bmatrix} \begin{bmatrix} \widehat{V}_{\text{in}} \\ \hat{d} \end{bmatrix} \tag{15.9}$$

$$Y(t) = \begin{bmatrix} 1 & 0 \\ 0 & 1 \end{bmatrix} \begin{bmatrix} \hat{i} \\ \hat{v} \end{bmatrix}$$

15.3.1 Continuous Time Transfer Functions of the System

It is depicted from the small signal model shown in Eq. (15.9), the inputs are $\hat{v}_g$ and $\hat{d}$ and the outputs are $\hat{v}_o$ and $\hat{i}_o$. The mathematical model of the DC–DC converter that gives relationship between the input and output variables can be assumed by Eqs. (15.10) and (15.11).

$$\hat{v}_o = G_{vg}\hat{v}_g + G_{vd}\hat{d} \tag{15.10}$$

$$\hat{i}_o = G_{ig}\hat{v}_g + G_{id}\hat{d} \tag{15.11}$$

Table 15.1 System parameters

Parameter	Value	Units
Input voltage	800	Volts
Output voltage	400	Volts
Output power	48 K	Watts
Switching frequency	100 K	Hz
Nominal voltage of the battery	320	Volts
Battery state of charge	50	Percent
Filter inductor	10.42 μ	Henry
Filter capacitor	120 μ	Farads

In order to get transfer function in Eqs. (15.10) and (15.11), each input one at each time is taken into consideration by equating other input to zero. Therefore, the transfer functions G_{vg}, G_{ig} are derived by Eqs. (15.12) and (15.13) when $\hat{d} = 0$.

$$G_{vg}(s) = \frac{D \times R_0}{\left(S^2 L_0 C_0 R_0 + S L_0 + R_0\right)} \tag{15.12}$$

$$G_{ig}(s) = \frac{D(1 + S C_0 R_0)}{\left(S^2 L_0 C_0 R_0 + S L_0 + R_0\right)} \tag{15.13}$$

Similarly, transfer functions G_{vd}, G_{id}, are given by (15.14) and (15.15) when $\widehat{V}_g = 0$

$$G_{vd}(s) = \frac{V_g \times R_0}{\left(S^2 L_0 C_0 R_0 + S L_0 + R_0\right)} \tag{15.14}$$

$$G_{id}(s) = \frac{V_g(1 + S C_0 R_0)}{\left(S^2 L_0 C_0 R_0 + S L_0 + R_0\right)} \tag{15.15}$$

In order to obtain the discrete-time model of the DC–DC converter from the above transfer function Eqs. (15.12)–(15.15), system parameters are calculated using desired quantities and are given in Table 15.1.

15.3.2 Discrete-Time Transfer Functions of the System

The conversion of the system into discrete-time domain model was found by means of pole–zero mapping technique with the operating sampling time equivalent to 5 μs and the transformed signal shown in Fig. 15.7.

The discrete-time model of converter is found using the methodology used in [7].

$$\hat{x}(n) = \Phi\hat{x}[n-1] + \Gamma\widehat{d_n}[n-1] \tag{15.16}$$

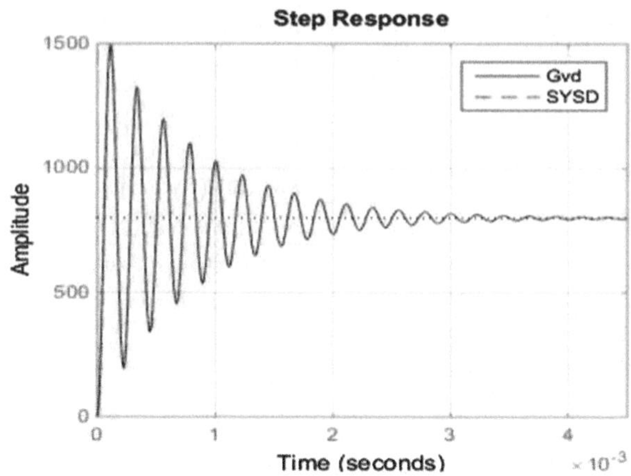

Fig. 15.7 Step response of continuous and discretized systems

where $\Phi = e^{A_1 D T_S} e^{A_2 D^1 T_S}$ & $\Gamma = \Phi K T s$

$$\Phi = e^{A T_S} = I + A T_S + \frac{A^2 T_S^2}{2} + \ldots$$

$$K = (A_1 - A_2)X + (B_1 - B_2)U$$

The small signal discrete-time model of converter is found by neglecting higher-order terms as

$$\Phi = \begin{bmatrix} 1 & -\frac{T_s}{L_0} \\ \frac{T_s}{L_0} & 1 - \frac{T_s}{R_0 C_0} \end{bmatrix} \tag{15.17}$$

$$\Gamma = \begin{bmatrix} T_s & -\frac{T_s^2}{2L_0} \\ \frac{T_s^2}{2C_0} & T_s - \frac{T_s^2}{2R_0 C_0} \end{bmatrix} \begin{bmatrix} \frac{d}{N L_0} & \frac{V_g}{N L_0} \\ 0 & 0 \end{bmatrix} = \begin{bmatrix} \frac{d T_S}{L_0} & \frac{V_g T_s}{L_0} \\ 0 & 0 \end{bmatrix} \tag{15.18}$$

Discrete-time transfer functions are determined using Eq. (15.19) and are presented in Eqs. (15.20)–(15.23),

$$T.F. = C(ZI - \Phi)^{-1}\Gamma + E \tag{15.19}$$

$$G_{ig}(Z) = \frac{\frac{dTs}{L_0}\left(Z - 1 + \frac{T_s}{R_0 C_0}\right)}{Z^2 - Z\left(2 - \frac{T_s}{R_0 C_0}\right) + \left(1 - \frac{T_s}{R_0 C_0} + \frac{T_s^2}{L_0 C_0}\right)} \tag{15.20}$$

$$G_{id}(Z) = \frac{\frac{V_g Ts}{L_0}\left(Z - 1 + \frac{T_s}{R_0 C_0}\right)}{Z^2 - Z\left(2 - \frac{T_s}{R_0 C_0}\right) + \left(1 - \frac{T_s}{R_0 C_0} + \frac{T_s^2}{L_0 C_0}\right)} \tag{15.21}$$

$$G_{vg}(Z) = \frac{\frac{dT_S}{L_0}\left(\frac{T_s}{C_0}\right)(z+1)}{Z^2 - Z\left(2 - \frac{T_s}{R_0 C_0}\right) + \left(1 - \frac{T_s}{R_0 C_0} + \frac{T_s^2}{L_0 C_0}\right)} \quad (15.22)$$

$$G_{vd}(Z) = \frac{\frac{V_g T_S}{L_0}\left(\frac{T_s}{C_0}\right)(z+1)}{Z^2 - Z\left(2 - \frac{T_s}{R_0 C_0}\right) + \left(1 - \frac{T_s}{R_0 C_0} + \frac{T_s^2}{L_0 C_0}\right)} \quad (15.23)$$

The appropriate design of the digital controller can be easily derived with the help of root locus technique from the intended design model of the discrete-time converter.

15.4 Controller Design

After successfully obtaining the most appropriate discrete-time modeling of the desired converter, root locus technique is used to design a controller for the considered system as shown in Fig. 15.8. It is observed that poor response, where the poles are located very near to the unit circle and it shows the system is at verge of instability and also has very poor 0.254 db phase margin as shown in Figs. 15.9 and 15.10. It necessitates the design of controller for getting desired output voltage.

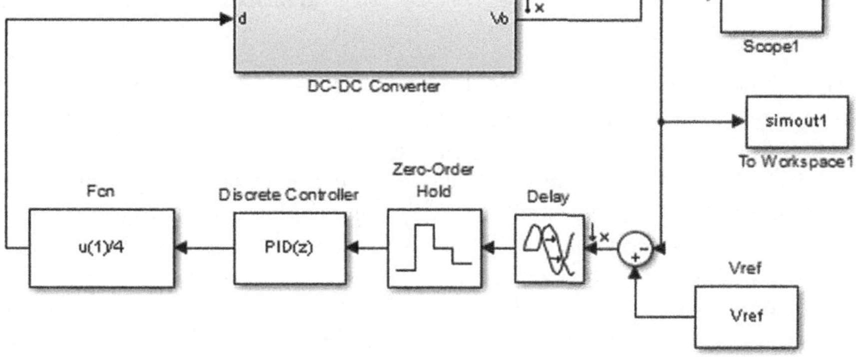

Fig. 15.8 DC–DC converter with digital controller

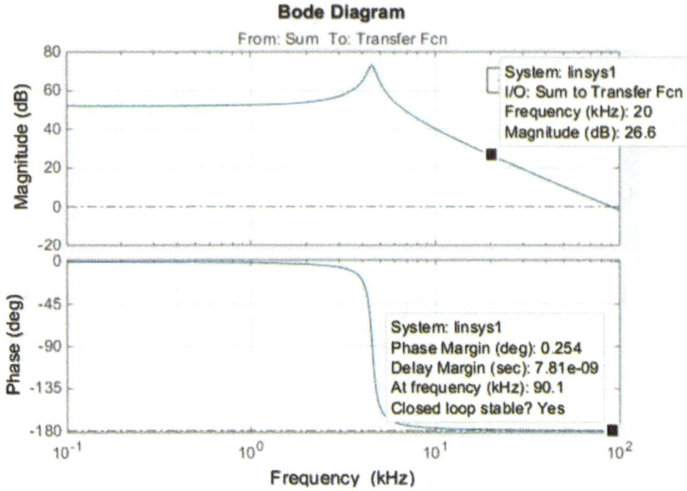

Fig. 15.9 Frequency response of the system without compensator

Fig. 15.10 Root locus of the system without compensator

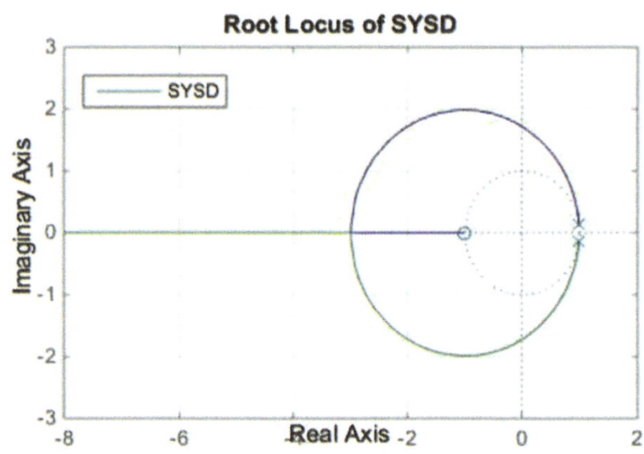

There are predominantly two procedures to develop a controller by implementing root locus method as designed by one of the authors in [5]. Where controller design is done with the help of direct digital design approach in MATLAB by placing pole and zeros and by adjusting the gain to get desired response. The other method is transforming the discrete-time transfer function into w-plane using Tustin method or using its equivalent relation $w = \frac{2}{T}\frac{z-1}{z+1}$ and then designs the controller with the help of frequency response method and again coming across root locus technique for the purpose of fine-tuning of gain of the controller. In this proposed work, direct digital design technique is used for the design of a controller by employing poles and zeros at the required locations with the help of root locus and frequency response of the system. To improve stability limit and phase margin along with reduction in steady-state error, a PID controller is designed to improvise the stability limit along with

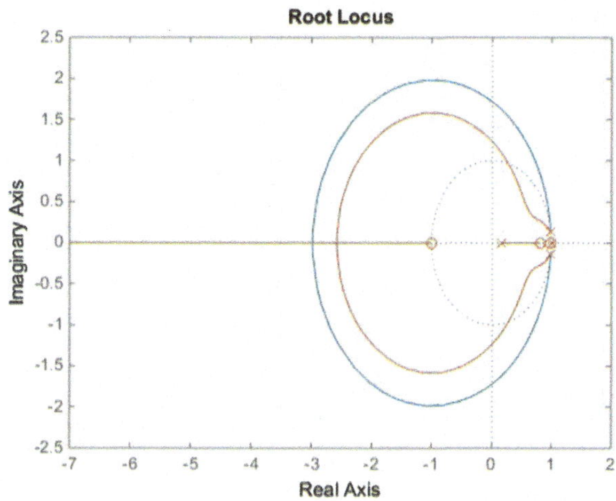

Fig. 15.11 Root locus of the system with compensator

reduction in steady-state error of the system by placing poles at $P = 1, 0.1446$ and zeros at $Z = 0.99, 0.8154$ and gain is so adjusted to 3.5475 to make phase margin at least 60° at 20 kHz. Figure 15.11 represents the system's root locus with the inclusion of designed controller. It assures the stability of the system as the response is inside the unit circle.

The transfer function of designed digital controller is represented in Eq. (15.24)

$$G_{cd}(z) = \frac{3.5475(z - 0.99)(z - 0.8154)}{(z - 1)(z - 0.1446)} \tag{15.24}$$

Even with the problem of sampling effect, digital controllers have a quite edge over existing analog controllers. They can be easily programmed to cater new design without changing the hardware of the system. After design of controller, further to analyze the validation of the proposed controller for the considered system, frequency domain and time domain analysis were made available with the results obtained from MATLAB/Simulink. Figure 15.12 illustrates the overview of the simulated circuit used for time domain analysis. The present work is concentrated only on design of controller for converter and hence corresponding results were presented.

Figure 15.13 shows Bode response of the system with controller with phase margin of 72.2° at desired crossover frequency 20 kHz, and Fig. 15.14 shows the output voltage of the system with different steps in voltages at 0.004 and 0.007 s with magnitude variation of ±25 V at input. Both the results validate the designed controller as the responses shown in Fig. 15.15 are very close to desired one.

The output voltage of DC–DC converter is dynamically varied with the change in the input source voltage, load rating and system parameters. Here in this application of battery charging, the load is considered as constant. So, the designed controller has to address change in voltage due to source and parametrial variation. Further to evaluate the performance of digital controller, different cases are considered in which

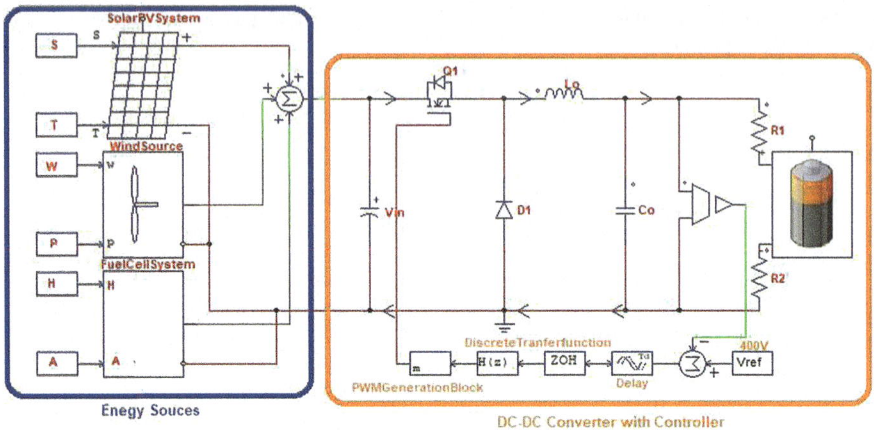

Fig. 15.12 Overview of the simulation circuit

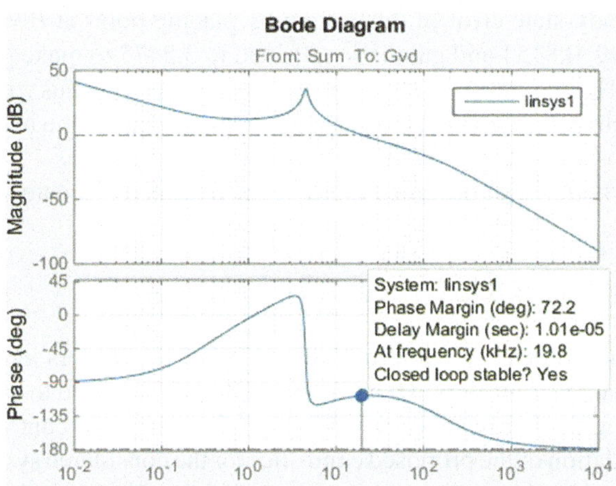

Fig. 15.13 Frequency response with compensator

one variation is input to the converter and the other one is variation in parameters of the converter as load remains same in this application. Figure 15.16 represents response of the system with designed controller for different input voltages 900 V (green), 850 V (blue), 750 V (red) and 700 V (sky blue). It assures the steady-state tracking with the designed controller.

Similarly, converter parameter is changed to different combinations Lo = 10.42 μH and 120 μF (green), 12 μH and 120 μF(blue), 10.42 μH and 150 μF(red) and 12 μH and 150 μF(sky blue) with same step in input voltages at 0.003 s. And results are presented in Fig. 15.17 though it assures the steady-state tracking capability.

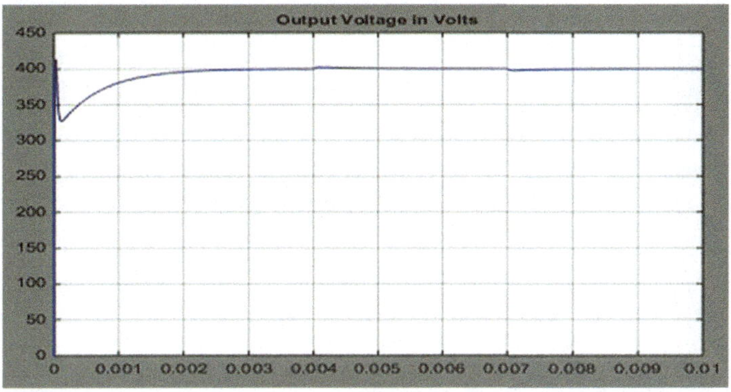

Fig. 15.14 Output voltage of the system with disturbances at 0.004 and at 0.007 s

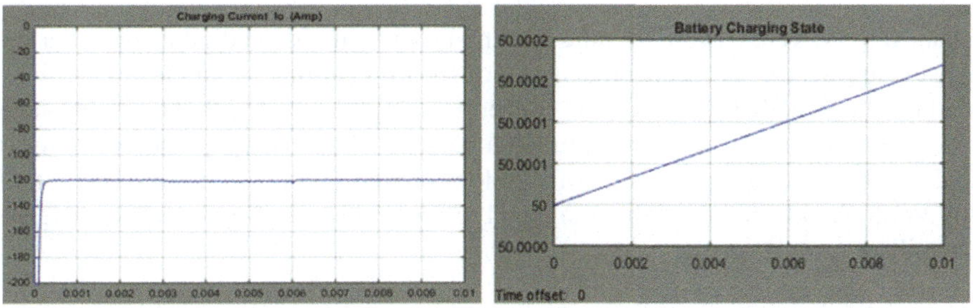

Fig. 15.15 Battery charging current and state of charge

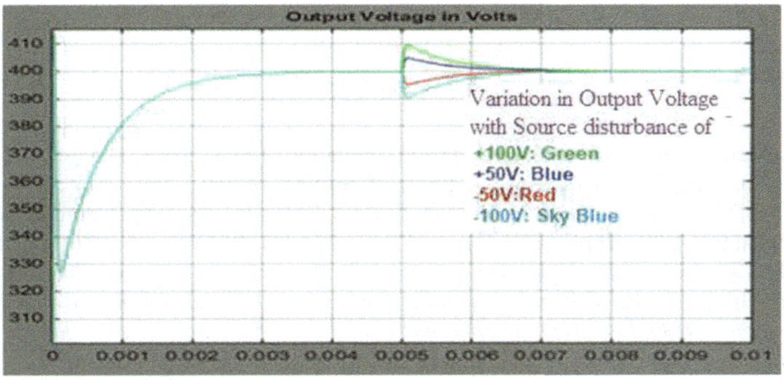

Fig. 15.16 Variation in output voltage with change in input

15.5 Conclusion

The proposal in this paper introduces a comprehensive design procedure to control the DC–DC converter with the help of a digital controller. The examination of the results

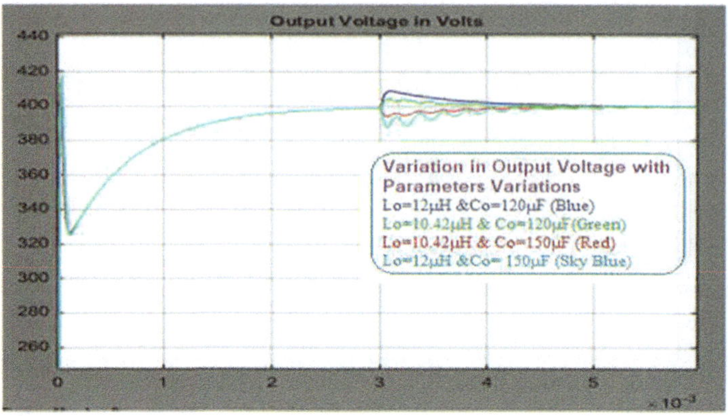

Fig. 15.17 Variation in output voltage with change in parameter values

obtained by MATLAB validates the design of root locus-based digital controller for a fast charging DC–DC converter in the off-board charging of an electric vehicle. Mathematical modeling of DC–DC converter is implemented with the help od state space averaging technique. The optimum operation of the used DC–DC converter is achieved by the proposed design of root locus-based digital controller. The output 400 V/120 A obtained from the designed DC–DC converter shows the fast charging capability.

Acknowledgements The authors are thankful to Bapatla Educational Society, Bapatla, for extending their support for this work.

References

1. G. Joos, M. de Freige, M. Dubois, Design and simulation of a fast charging station for PHEV/EV batteries, in *IEEE Electrical Power & Energy Conference* (2010)
2. G.R. Chandra Mouli, P. Bauer, M. Zeman, System design for a solar powered electric vehicle charging station for workplaces, in *Applied Energy*, vol. 168 (Elsevier, 2016), pp. 434–443
3. H. Li, H. Liu, A. Ji, F. Li, Y. Jia, Design of a hybrid solar-wind powered charging station for electric vehicles, in *IEEE International Conference on Materials for Renewable Energy and Environment, Chengdu* (2013), pp. 977–981
4. A. Verma, B. Singh, An implementation of renewable energy based grid interactive charging station, in *IEEE Transportation Electrification Conference and Expo (ITEC), Detroit* (2019)
5. H. Fathabadi, Novel solar powered electric vehicle charging station with the capability of vehicle-to-grid. Solar Energy **142**, 136–143 (2017)
6. N. BalaKrishna, B. Santhana Krishnan, R. Janga, Performance evaluation of constant current controllers used in fast charging of electric vehicle with fuel cell system. J. Adv. Res. Dyn. Control Syst. **12**(02), 2205–2215 (2020)
7. R. Janga, S. Malaji, Root locus Based design of digital controller for forward converter with active clamp circuit. Int. J. Appl. Eng. Res. **10**(1), 20–25 (2015)
8. R.W. Erickson, D. Maksimovic, *Fundamentals of Power Electronics*, 2nd edn (Springer publications, 2001)

9. R. Janga, S. Malaji, Digitally controlled active clamp forward converter with small signal discrete-time modeling, in *IEEE Conference Proceedings ICCCi* (2014), 1–6
10. V.T. Tran, K.M. Muttaqi, D. Sutanto, A robust power management strategy with multi-mode control features for an integrated PV and energy storage system to take the advantage of ToU electricity pricing, in *IEEE Transaction on Industry Application*, vol. 55, no. 2, March-April 2019, pp. 2110–2120
11. M.F. Shaaban, S. Mohamed, M. Ismail, K. Qaraqe, E. Serpedin, Joint planning of smart EV charging stations and DGs in eco-friendly remote hybrid microgrids, in *IEEE Transaction on Smart Grid, Early Access*

Chapter 16
High-Voltage Electric Water Pump for the Application of Electric Vehicle—Modeling, Design and Analysis

Aneesh Jategaonkar and N. C. Lenin

Abstract This paper presents the designing and electromagnetic analysis of permanent magnet brushless DC (PMBLDC) motor using finite element analysis software packages for high-voltage water pump application. The designed water pump motor is suitable for application in the battery pack cooling system used in electric vehicles (EVs). Two PMBLDC models based on 200 volt system and 300 volt system are proposed for the water pump application. Component-wise designing procedure is explained for the PMBLDC motor to obtain efficiency improvement and cost reduction compared to the prominently used models in the industry. Elaborate study on performance estimation is carried out for both the models using electromagnetic analysis.

Keywords PMBLDC motor · Finite element analysis · Electromagnetic analysis · Water pump · Electric vehicle

16.1 Introduction

Cooling systems are essential in electric vehicles as heat is continuously generated in the battery pack. If the generated heat is left unattended to, it may lead to serious problems like excessive heat generation ushering to fire. Performance of the cells in the battery pack is also affected in a grave manner pertaining to the excessive heat generation [1]. The electric vehicle market is expanding rapidly, and studies indicate that globally it will reach $802.81 billion by the year 2027 with a compound annual growth rate of 22.6% [2]. To meet the excessive demand for cooling system of the battery pack that would be generated by the swelling EV market, it is necessary to come up with high-efficiency, low-cost and reliable water pump motors.

PMBLDC motors prove to be superior as compared to other traditionally used machines like induction motors for the water pump application in an EV. PMBLDC motors are compact as compared to induction motors with similar ratings due to the

A. Jategaonkar · N. C. Lenin (✉)
Vellore Institute of Technology, Chennai Campus, Chennai 600127, India
e-mail: lenin.nc@vit.ac.in

© The Author(s), under exclusive license to Springer Nature Singapore Pte Ltd. 2021
V. Kamaraj et al. (eds.), *Emerging Solutions for e-Mobility and Smart Grids*,
Springer Proceedings in Energy, https://doi.org/10.1007/978-981-16-0719-6_16

use of permanent magnets, thus eliminating the need for armature winding. Weight of the motor also can be reduced significantly as compared to the induction motor due to the elimination of armature winding [3]. The cost for controller unit can be saved on using a PMBLDC machine as the hardware required for PWM control is cheaper than the hardware required for frequency control of an induction motor [4].

Compactness offered by the PMBLDC machine is critical due to space restrictions in an EV. The use of liquid coolants like water is superior as compared to air cooling due to higher heat conductivity and heat capacity than air. Advantages like compact structure and ease of arrangement are also offered by the fluid cooling system [5].

Performance estimation using electromagnetic analysis is an extremely critical step before going for hardware manufacturing. This is so, as it saves a lot of time and money which would have been spent if the hardware prototype was developed initially. Finite element analysis software packages help to provide the platform for this software modeling and analysis [6]. Important parameters like output power, torque and speed are to be met while the modeling stage maintaining the efficiency high.

As the electric vehicle has a battery pack for supplying the drive motor, thus sufficient DC voltage is readily available for the water pump motor also. In hybrid or plug-in hybrid vehicles, the battery pack voltage is typically around 100–200 volt, whereas this voltage typically lies in the range of 400–800 V for electric only vehicles [7]. Thus, two PMBLDC models operating on voltage level of 200 volt and 300 volt are proposed in this paper for use in hybrid/plug-in hybrid and pure electric vehicles respectively. 8-pole, 12-slot models are developed using finite element software packages for both the voltage systems. The pack size for both the models is kept the same, thus eliminating any sizing constraints while interchanging the models for replacement when supply voltage restrictions are not applicable.

All the necessary specifications are stated in Sect. 16.2. Section 16.3 deals with the designing procedure followed. Stator, winding and rotor designs are explained elaborately in Sects. 16.3.1, 16.3.2 and 16.3.3, respectively. Sections 16.4.1 and 16.4.2 explain the electromagnetic performance estimation results obtained for the 200 volt- and 300 volt-based model, respectively. Cost analysis for each of the models is carried out in Sects. 16.5.1 and 16.5.2. The conclusions which were drawn from the entire study are presented in Sect. 16.6. Ansys Maxwell was chosen as the finite element software for designing and carrying the performance estimation of the motors [8]. Figure 16.1 shows the 200 volt PMBLDC water pump motor which was developed for the battery pack cooling application. Important components of the motor are labeled for clarity.

16.2 Specifications

The necessary characteristics required for the water pump cooling motor is high output power sufficient to be able to generate the force for pumping water. The rated speed in thousands of rpm ensures that water movement is speedy for timely and

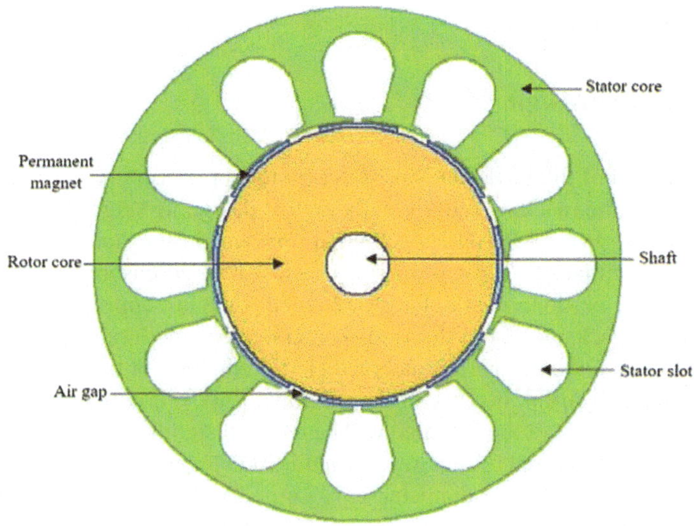

Fig. 16.1 PMBLDC motor for water pump application

efficient cooling. Torque rating is also important to ensure that the load demand is driven appropriately [9]. Table 16.1 gives the important specifications which were considered before initiating the design of the PMBLDC motor. The design was initiated to meet these parameters at high efficiency and low cost. All the required specifications for the 200 volt and 300 volt system are the same except the supply voltage.

Table 16.1 Specifications

Quantity	Required numerical value
Output power	380 W
Speed	4600 rpm
Torque	0.78 Nm
Supply voltage	200 volt and 300 volt DC
Stack length	72 mm
Stator outer diameter	127 mm
Stator slot fill factor	≤45%
Shaft diameter	16 mm
Stator core steel type	M49_29G
Rotor core steel type	M49_29G
Winding material	Copper
Permanent magnet material	Ceramic (ferrite)

16.3 Design Approach

16.3.1 Stator Design

The stator design of the PMBLDC for water pump application is a critical stage as the durability of the machine relies greatly on this designing stage. This is so as the slots for winding in a PMBLDC machine are present in the stator. The modeling is done to ensure that fill factor is restricted below 45% to ensure simplicity while the winding will be done on the hardware stage. Leaving appropriate stator core thickness at the backside of the slot is also crucial to ensure durability of the design as the sturdiness of the design is concerned [10]. Reduction of the size of the stator outer diameter will shrink the size of the machine. This will optimize the design to great extents as weight of the machine will be reduced significantly due to reduction in the weights of stator and rotor cores. Both the 200 volt and 300 volt systems were made keeping the stator outer diameter as 127 mm. The steel type used in the stator core of 200 volt and 300 volt system was M49_29G. The stacking factor of both the machines was chosen to be 0.95 considering the feasibility and practicality of the design constraints. Figure 16.2 shows the slot design type which was chosen for modeling the stator. Rounding of slots is important in the design to ensure that flux lines do not break while operation. Making the sides of adjacent slots parallel to each other makes the physical procedure of winding easy while manufacturing the PMBLDC motor [11].

The stator slot parameters which were fixed for the 200 volt PMBLDC system are shown in Table 16.2. These parameters were optimized for fill factor and stator core weight minimization. The parameter Hs2 aids in the lengthening of the stator slot so as to increase side length facing the adjacent slots to ensure ease while winding physically. The parameters Bs1 and Bs2 increase the thickness of the slot to provide

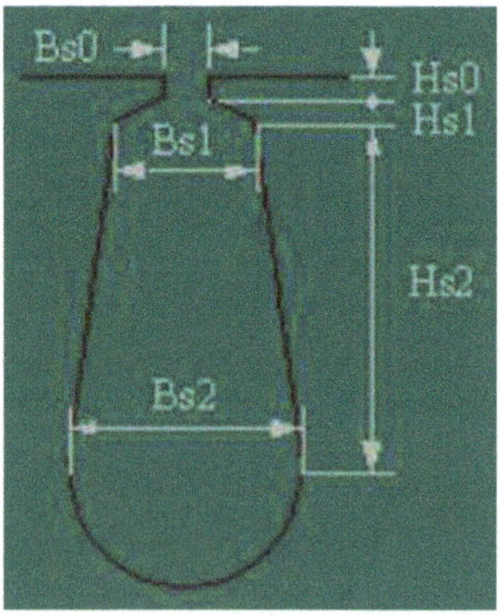

Fig. 16.2 Stator slot design type chosen for 200 volt and 300 volt PMBLDC water pump motor

Table 16.2 Slot configuration used in 200 volt PMBLDC system

Parameters	Numerical value (mm)
Hs0	0.5
Hs1	0.5
Hs2	12
Bs0	2.3
Bs1	12
Bs2	18

Table 16.3 Slot configuration used in 300 volt PMBLDC system

Parameters	Numerical value (mm)
Hs0	0.5
Hs1	0.5
Hs2	12
Bs0	2.2
Bs1	12
Bs2	18

the space needed for windings. The analytical design of the PMBLDC motor was also considered prior to the initiation of the designing procedure [12].

Table 16.3 shows the slot configuration parameters which were chosen for the 300 volt PMBLDC motor. The only important difference between the 200 volt and 300 volt system which was made while designing the slots was that the length of Bs0 was decreased in the 300 volt motor as compared to the 200 volt one. This change was made to accommodate the changes in the windings which exist between both the models. Correct slot design for each and every design according to the expected performance is important as it helps to reduce the unwanted cogging torque which may get amplified on implementing incorrect stator slot design. If slot design according to desired characteristics is neglected, it may lead to a sharp fall in the control precision [13].

16.3.2 Winding Design

The whole coiled winding configuration is used in both the PMBLDC designs. A double-layered winding configuration was implemented in both the designs. In this type of configuration, the stator slot is divided into two parts, namely 'top' and 'bottom.' The coils are wound from the bottom part of one slot to the top part of the adjacent slot. The number of parallel branches, coil pitch and the number of strands for both the designs were kept one. The back electromotive force obtained due to this winding design is trapezoidal in nature, as is the characteristic of brushless DC motors [14]. The number of conductors per slot parameter is used to fix the number

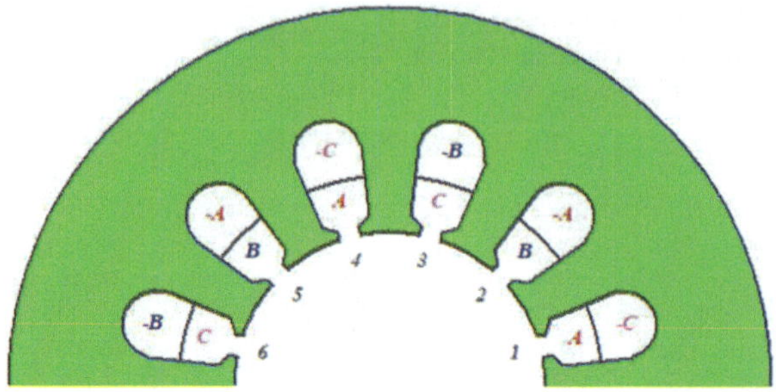

Fig. 16.3 Winding configuration

of turns of the winding in the design. The number of conductors per slot was fixed as 89 in the PMBLDC design based on 200 volt, while it was kept as 134 in the 300 volt system. Figure 16.3 shows the double-layered whole coiled winding configuration. A, B and C indicate the phases and the negative signs before the letters indicate the other sides of the coils.

16.3.3 Rotor Design

The rotor in a PMBLDC machine has permanent magnets in place of winding which is present in a rotor of a conventional DC motor. The thickness of the permanent magnets is fixed as per the performance requirements and required efficiency improvement. Low-cost ferrite-based magnet is implemented in both the designs considering cost optimization. The steel type which is used for rotor core is same as that of the stator core which is M43_29G. The stator and rotor lengths are kept same to keep the stack length of both the designs uniform. Thickness of the magnet is kept as 1 mm in both the designs. It is much less as compared to the commonly used magnet thickness of 6 mm in PMBLDC machines. The magnet embrace is kept as 0.7, which means that magnet occupies 70% of the outer circumference of the rotor. Shaft diameter is also specified while specifying the rotor dimensions as shaft diameter is effectively the inner diameter of the rotor.

16.4 Electromagnetic Results

16.4.1 PMBLDC Motor Based on 200 Volt

All the required performance characteristics of the PMBLDC motor for the water pump application were obtained for the 200 volt-based system. Efficiency of 93.97% was obtained at the rated parameters. Entire electromagnetic performance result of the machine is shown in Table 16.4. It can be observed from Table 16.4 that the output

Table 16.4 200 volt water pump motor performance

Parameters	Numerical value
Average input current (A)	2.06
Armature current density (A/mm^2)	1.85
Frictional and windage loss (W)	3.8
Iron core loss (W)	10.5
Armature copper loss (W)	7.88
Total loss (W)	24.92
Output power (W)	388.59
Input power (W)	413.52
Efficiency (%)	93.97
Rated speed (rpm)	4600
Rated torque (N m)	0.80
Stator slot fill factor (%)	43.04
Ideal back-emf constant (V s/rad)	0.29

power obtained is 388.59 W. This satisfies the requirement of 380 W as required by the water pump application. Major constituent of the loss which takes place in the motor is the iron core loss. These are basically the losses which take place due to the eddy current and hysteresis losses [15]. The total iron core loss of 10.5 W takes place in the 200 volt motor as can be seen from Table 16.4. The stator slot fill factor obtained for the design is 43.04% which ensures feasibility of the created design. Thus, the winding can be done without any special requirements of winding machinery. Figures 16.4, 16.5 and 16.6 portray the important characteristics of the 200 volt-based PMBLDC motor. It can be seen from Fig. 16.4 that the locked rotor current is 176 A. Although this condition never arises practically while the motor is in operation, current protection circuitry is provided to take care of any surges or overcurrent. It can be also seen from Fig. 16.4 that the no load speed of the machine is 6314 rpm. The locked rotor torque is 10.13 N m as can be seen from Fig. 16.5. Torque versus input current variation is shown in Fig. 16.6. Torque rises with input current till the value of torque becomes 9.4 N m and then torque becomes almost constant with rise in current. Operating characteristics of the motor can be decided

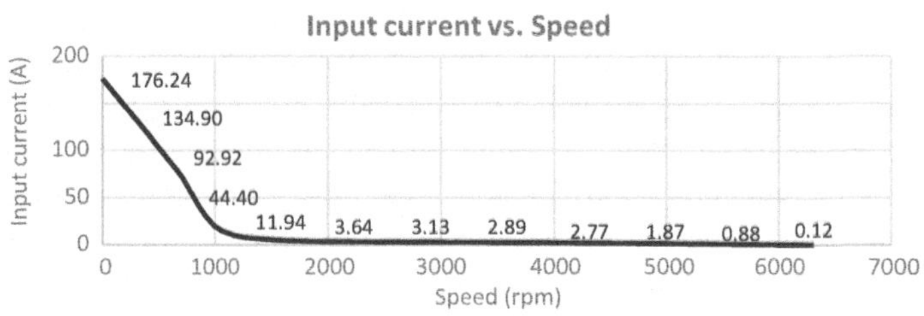

Fig. 16.4 Input current versus speed

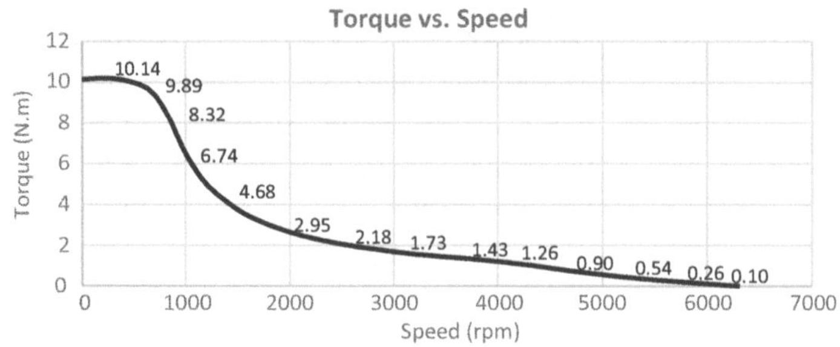

Fig. 16.5 Torque versus speed

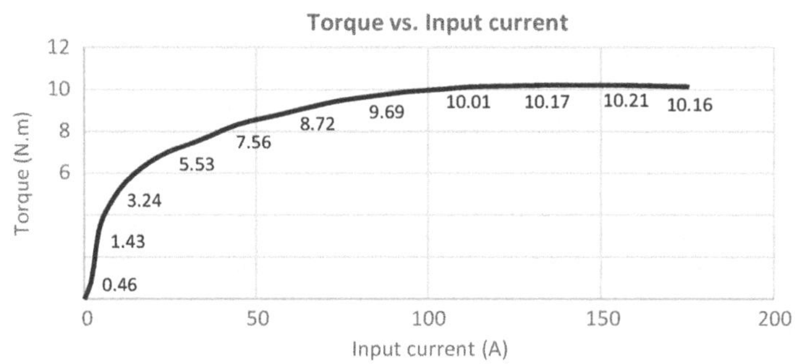

Fig. 16.6 Torque versus input current

as per the requirements on observing the described graphs in detail. Performance of the machine sways a bit by deflecting from the rated parameters during actual operation. Thus considering the operating characteristic curves of the PMBLDC motor is important [16].

16.4.2 PMBLDC Motor Based on 300 Volt

Extensive performance estimation of the 300 volt PMBLDC motor is represented in Table 16.5. It can be observed from the below table that the value of average input current is less as compared to the 200 volt PMBLDC design. This decrease in average input current can be attributed to the increase in the value of supply voltage with output power being about the same. Efficiency obtained in both the models is about the same which is approximately 94%. The major constituent in the loss incurred in the motor is the iron core loss. The value for the iron core loss incurred in the 300 volt design is around 10.53 W. The total loss suffered in the 300 volt design is around 24.06 W as can be observed from Table 16.5. All the necessary output parameters like output power, speed and torque were obtained when the stator slot

Table 16.5 300 volt water pump motor performance

Parameters	Numerical value
Average input current (A)	1.34
Armature current density (A/mm^2)	1.91
Frictional and windage loss (W)	3.8
Iron core loss (W)	10.53
Armature copper loss (W)	7.94
Total loss (W)	24.06
Output power (W)	380.08
Input power (W)	404.14
Efficiency (%)	94.04
Rated speed (rpm)	4600
Rated torque (N m)	0.78
Stator slot fill factor (%)	40.61
Ideal back-emf constant (V s/rad)	0.44

fill factor was 40.61%. This ensures that winding will be easy and special machinery will not be required even during the physical prototyping stage. Figures 16.7, 16.8 and 16.9 depict the important operating curves of the 300 volt-based PMBLDC motor.

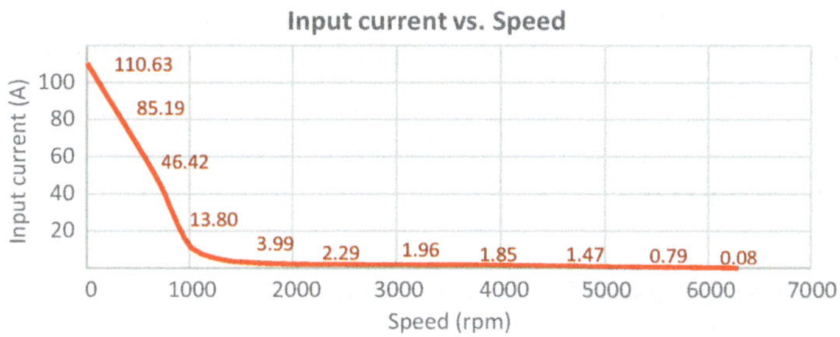

Fig. 16.7 Input current versus speed

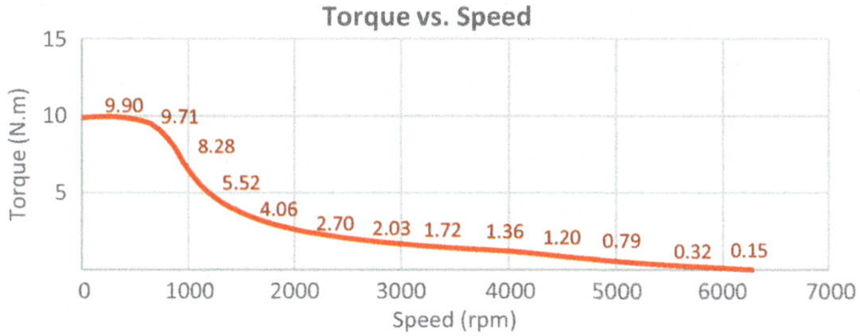

Fig. 16.8 Torque versus speed

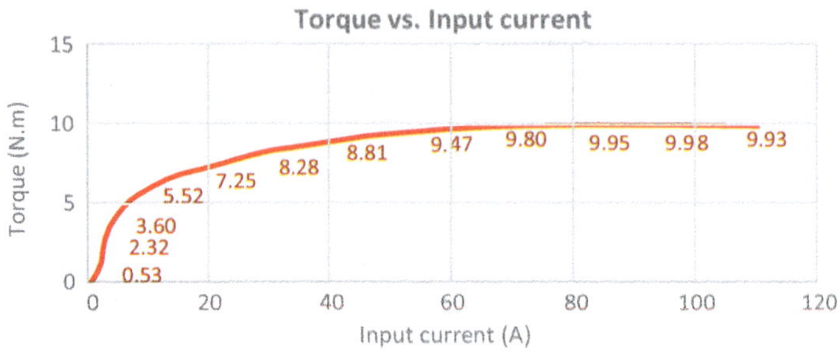

Fig. 16.9 Torque versus input current

The locked rotor current of the 300 volt-based PMBLDC motor is around 110 A, as is evident from Fig. 16.7; although this value will never be reached during actual operation as current controlling techniques are existent and the current will be limited at a much lower level. It can be observed from Figs. 16.8 and 16.9 that the locked rotor torque is around 9.8 N m for the 300 volt design. Value of torque increases as input current is increased till the value of 9.7 N m, but it becomes almost constant as the 9.7 N m threshold is crossed. As the machine is constructed by considering the rated parameters, it is meant for long lasting operating when used at the rated parameters and conditions. However, to comply with uncontrollable circumstances, the motor design is made so that it will sustain a current surge whose value will be 1.5 times that of the rated current value. All this kind of performance estimation is done by simulating the electromagnetic analysis using finite element software packages.

16.5 Cost Analysis

16.5.1 200 Volt-Based PMBLDC Motor

Table 16.6 gives the component-wise cost analysis for 380 W water pump with input voltage supply of 200 volt. Typically available 380 W PMBLDC motors for water

Table 16.6 Cost analysis for 200 volt-based PMBLDC motor

Components	Weight (g)	Price per kg (rupees)	Cost (rupees)
Winding	944.46	900	849.6
Permanent magnet	53.53	505	27.03
Core weight (stator + rotor)	4393.09	200	878.618
Total cost			1755.24

Table 16.7 Cost analysis for 300 volt-based PMBLDC motor

Components	Weight (g)	Price per kg (rupees)	Cost (rupees)
Winding	887.80	900	799.02
Permanent magnet	53.53	505	27.03
Core weight (stator + rotor)	4393.56	200	878.71
Total cost			1704.76

pump in the market cost around 5000 rupees. Thus, it can be observed that significant cost saving can be achieved on using the proposed model. This cost saving is achieved primarily due to the use of cheaply available ferrite-based magnets. Typically, the expensive NdFeB-based magnets are used in industries when permanent magnet-based motors are required, thus increasing the permanent magnet cost at least ten times as compared to the ferrite magnet [17].

16.5.2 300 Volt-Based PMBLDC Motor

The constituent-wise cost analysis of the 380 W PMBLDC water pump motor with input voltage of 300 volt is given in Table 16.7. The winding weight is reduced in the 300 volt design, but the core weight is about the same as compared to the 200 volt design. As the cost of copper in the windings is less than the 200 V motor, the overall cost of the designed model is slightly less than the 200 volt model cost.

16.6 Conclusion

PMBLDC motors were successfully designed, and the performance estimation was carried out for both the 200 volt and 300 volt input voltage system using finite element software packages. Both the motors operate at 94% efficiency at rated conditions which is higher than the standard operating efficiency. Cost saving was achieved in both the motor designs as compared to the commonly used water pump motors in the industry due to the usage of inexpensive ferrite-based magnets. Based on the study done, the following inferences are made by the authors:

- Reduction in average current of about 0.72 A is obtained on implementing the 300 volt-based motor as compared to the 200 volt motor.
- The ideal back-emf constant of the 200 volt design is 0.15 volt-second per radian less than the 300 volt design.
- There is a slight difference in efficiency of 0.074% between the designed 200 volt- and 300 volt-based models.

- Total net weight of the 300 volt-based model is less than 56.19 grams as compared to the 200 volt-based model.

References

1. Z. Chong, L. Fei, Z. Hua, S. Jing, M. Chris, A real-time battery thermal management strategy for connected and automated hybrid electric vehicles (CAHEVs) based on iterative dynamic programming. IEEE Trans. Veh. Technol. **67**(9), 8077–8084 (2018)
2. https://www.alliedmarketresearch.com/electric-vehicle-market
3. Z. Qinghu, C. Siwei, W. Dong, J. Zhewu, Multiobjective design optimization of high-power circular winding brushless DC motor. IEEE Trans. Ind. Electron. **65**(2), 1740–1750 (2018)
4. V. Rodolfo, A. Pedro, F. Andre, B. Pedro, Unipolar PWM predictive current-mode control of a variable-speed low inductance BLDC motor drive. IET Electr. Power Appl. **11**(5), 688–696 (2017)
5. R. Rania, L. Hasna, G. Hamid, S. Pierre, Passive cooling of high capacity lithium-ion batteries, in *IEEE International Telecommunications Energy Conference 2018* (IEEE, Turin, 2018), pp. 1–4
6. S. Toshihito, S. Yuki, S. Takahiro, I. Hajime, Fast finite-element analysis of motors using block model order reduction. IEEE Trans. Magn. **52**(3), 1–4 (2016)
7. G. Xianzhi, X. Rui, M. Chunting, Study of the characteristics of battery packs in electric vehicles with parallel-connected lithium-ion battery cells. IEEE Trans. Ind. Appl. **51**(2), 1872–1879 (2015)
8. https://www.ansys.com/products/electronics/ansys-maxwell
9. P. Joon, C. Jun, G. Bon, J. In, BLDC drive control of electric water pump for automotive application, in *IEEE Vehicle Power and Propulsion Conference 2010* (IEEE, Lille, 2010), pp. 1–5
10. K. Hoe, J. Tae, Design of BLDC motor for water pump using bonded-ferrite, in *IEEE 2013 Tencon—Spring* (IEEE, Sydney, 2013), pp. 199–203
11. J. Mehrdad, F. Hossein, Optimum design of the stator parameters for noise and vibration reduction in BLDC motor. IET Electr. Power Appl. **12**(9), 1297–1305 (2018)
12. P. Kumar, P. Bauer, Improved analytical model of a permanent-magnet brushless DC motor. IEEE Trans. Magn. **44**(10), 2299–2309 (2008)
13. K. Rajan, S. Bhim, Grid Interactive solar PV-based water pumping using BLDC motor drive. IEEE Trans. Ind. Appl. **55**(5), 5153–5165 (2019)
14. C. Xi, L. Gang, Sensorless optimal commutation steady speed control method for a nonideal back-EMF BLDC motor drive system including Buck converter. IEEE Trans. Ind. Electron. **67**(7), 6147–6157 (2020)
15. P. Yeji, K. Hyunwoo, J. Hyungkwan, H. Sang, L. Ju, J. Dong, Efficiency improvement of permanent magnet BLDC with Halbach magnet array for drone. IEEE Trans. Appl. Supercond. **30**(4), 1–5 (2020)
16. S. Aryadip, S. Bhim, Peak current detection starting based position sensorless control of BLDC motor drive for PV array fed irrigation pump, in *IEEE International Con-ference on Environment and Electrical Engineering and IEEE Industrial and Commercial Power Systems Europe (EEEIC/I&CPS Europe)* (IEEE, Italy, 2019), pp. 1–6
17. S. Sashidhar, G. Fernandes, A Novel Ferrite SMDS spoke-type BLDC motor for PV bore-well submersible water pumps. IEEE Trans. Industr. Electron. **64**(1), 104–114 (2017)

Chapter 17
Charging and Discharging Characterization of a Community Electric Vehicle Batteries

D. Suganthi and K. Jamuna

Abstract Lithium-ion batteries (LIBs) play a major role in the electrified world which needs the technological innovations. Many electric vehicles and hybrid electric vehicles are fortified with LIB that extenuates environmental pollution and reduces the usage of the conventional energy storage devices. This paper outlines the charging and discharging characteristics of a single LIB and multiple LIBs present in the EVs located in the community and also calculates the battery power. Mathematical equations are formed based on the chemical behaviour of LIB. The simulations are performed in MATLAB, and the results are discussed.

Keywords Lithium-ion battery · Electric vehicle · SoC · Charging and discharging characteristics

17.1 Introduction

Electric vehicle (EV) reduces fossil fuel, greenhouse gas emission, and clean transportation around the world. Nowadays, EVs development is mainly focused on renewable energy. Electric vehicles' share is expected to increase from 2 to 22% in 2030. Among the storage devices, the battery is an important device for EVs. Compared to other batteries, the lithium-ion battery (LIB) has several benefits such as longer cycle life, high energy density, fast charging/discharging, safety, and no memory effect. EVs act as controllable loads in demand response programs [1] in a power system network. Another interesting application is that EVs actively interact with the electrical grid as an energy bank for the renewable power sources connected power networks popularly defined as V2G [2]. Different types of battery models present in this literature [3] are electrochemical models [4], mathematical models, and equivalent circuit models. For the perfect simulation of EVs, HEVs, and PHEVs, it is necessary for an accurate battery model which predicts the battery characteristics fast and convenient [5]. Battery electric vehicles (BEVs) are optimized for

D. Suganthi (✉) · K. Jamuna
Vellore Institute of Technology Chennai, Chennai, India
e-mail: suganthi.d2019@vitstudent.ac.in

© The Author(s), under exclusive license to Springer Nature Singapore Pte Ltd. 2021
V. Kamaraj et al. (eds.), *Emerging Solutions for e-Mobility and Smart Grids*, Springer Proceedings in Energy, https://doi.org/10.1007/978-981-16-0719-6_17

various operating ranges, for example, driving range (i.e. the maximum distance an EV runs when it was fully charged). Hybrid electric vehicles (HEVs) price range is similar to the internal combustion engine (ICE) vehicles, and the fuel usage is 40 to 60% of ICE vehicles. The major components of the EVs are batteries. EV models are mainly electrochemical model which is influenced by the chemical reactions and define the features of battery and its action [6]. LIB is a typical electrochemical energy storage system. The Thevenin's RC model of a LIB consists of an RC dynamic network, internal resistance, open-circuit voltage, and the terminal voltage [7]. The Thevenin's equivalent model of Li-ion battery improves accuracy, and it is used to calculate the available power [8].

An optimization problem is framed for the cost savings and the reduction of grid emission with the utilization of electric vehicles and renewable energy sources [9, 10]. EVs are worked as either load demand or source in a smart grid or microgrid systems. For safety purpose and reliable operation of the battery, SoC should be determined [11], and the methods are direct voltage measurement, impedance measurement, adaptive methods like Kalman filtering, extended Kalman filtering, and artificial neural network [12].

The paper is structured as follows: in Sect. 17.1.2, Thevenin's equivalent circuit model of the LIB battery and major battery parameters are discussed. The chemical reaction of the LIB is also presented. In Sect. 17.2, the paper mainly focuses on the charging and discharging characteristics of LIB which is embedded in the electric vehicles. The characteristics will be useful for further applications like V2G and G2V. Hence, the electric vehicles located in the society or community's parking lot could be used for the above applications. These electric vehicles are denoted as the community's electric vehicles. EVs charge its battery from the grid and discharge the stored energy back to the grid. The energy exchange is being performed as G2V and V2G through the EV charging system. In order to realize the charging system concepts of EVs, single EV battery charging and discharging process is performed. While doing meaningful service to the grid, many EV's batteries should be synchronized and meet the grid's demand when it is required. The simulation results of a single battery and multiple batteries charging/discharging characteristics are presented and discussed in Sect. 17.3. Finally, the conclusion is given in Sect. 17.4.

17.1.1 Thevenin's Equivalent Model of LIB

The model uses on voltage source to represent the open-circuit voltage (OCV) and the polarization resistance R_1 with an internal resistance R_0 as stated in [13]. The Thevenin's battery model is shown in Fig. 17.1.

The battery terminal voltage is calculated as follows

$$V_B = E_m - R_0 i_t - R_1 i_t \tag{17.1}$$

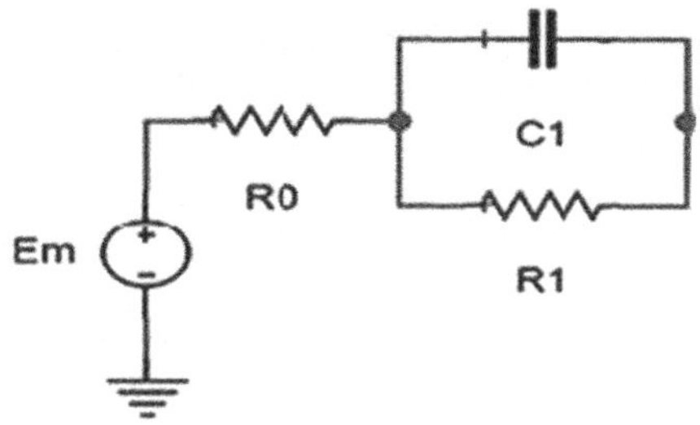

Fig. 17.1 Thevenin's RC circuit model

where

V_B	Battery terminal voltage
E_m	Open-circuit voltage of battery
i_t	Operating current-charging (positive) and discharging (negative)
R_0	Ohmic resistance of the battery
R_1	Polarization resistance

The battery model is used to measure battery values like battery terminal voltage, current, and its capacity. A battery terminal voltage is a gauge to judge that the battery is fully discharged/charged or not. The battery difficulties and accuracy will affect the estimation result of the battery SoC, capacity, and SoH. In this paper, the open-circuit voltage method is used for estimating the SoC.

The electrochemical model describes the characteristics of the battery cell which relates the correlation among SoC and temperature. The complexity of this model needs more computing powers and monitoring algorithms. Battery models could be designed with artificial neural networks, fuzzy logic, and fuzzy created neural networks. The electric model method is very popular in battery since it produces more accurate values. The SoC is the direct relationship of open-circuit voltage in an electric model. Its dynamic characteristics and high accuracy make this method widely used in all different kinds of electric vehicle applications. Here, the OCV value is considered to be a fixed value. For simplicity and real-time purpose, the Thevenin's RC method is used in this paper.

17.1.2 Major Battery Parameters

State of charge (SoC) is defined as the ratio of the available capacity to the rated capacity at a certain discharge rate [14]. The accuracy of the estimated SoC value has a great value in electric vehicular technology. Estimation of SoC prevents the battery from overheating, over-charging/discharging which increase the lifetime of the battery and ensured safety during its use. The SoC formula is as follows:

$$\mathrm{SoC} = \frac{Q_c}{Q_n} \times 100\% \tag{17.2}$$

where Q_c the remaining capacity, Q_n is the rated capacity and SoC is expressed as in percentage, $Q(I_n)$ is the discharge current. The SoC is also expressed as follows

$$\mathrm{SoC} = [Q_n - Q(I_n)]/Q_n \tag{17.3}$$

$$Q(I_n) = \int_{t_0}^{t} I_n \mathrm{d}t \tag{17.4}$$

SoC estimation is an important task in the batteries which could be performed in different methods, namely ampere-hour counting, adaptive filter method, and open-circuit voltage-related SoC estimation [15]. The battery terminal voltage is also known as OCV. Many papers on cell modelling do not directly consider the estimating SoC, and on estimating the SoC includes the cell modelling first and it does not depend on battery's age and temperature [15]. In this paper, the open-circuit voltage is assumed as 45 V. DoD is discharged electrical energy divided by the total stored energy.

C-rate is rate at which the capacity of battery able to operate effectively. The *C*-rate is defined as in Eq. (17.5).

$$I_b = C/N \tag{17.5}$$

where C is the capacity (Ah), N is the number of hours of discharge, and I_b is the given battery current. For example, a 20-Ah battery and 10 h discharge means $b =$ 20 Ah/10 h = 2 A. This means a 20-Ah battery can supply 2 A average up to 10 h. Due to losses, a 20 Ah battery can supply 2 A average for less than 10 h. Here, the C-rate is used as C/0.5. The capacity of a battery may be higher than its normal rating, and it is possible for the depth of discharge value to exceed the nominal value. DoD means the percentage of the battery that has been discharged relative to the available capacity of the battery. For example, Tesla Powerwall holds 13.5 kilowatt-hours (kWh) of electricity, discharge is 13 kWh, and the DoD is approximately 96%. The cycle life is defined as the number of times a battery charge and discharge before it reaches its end life.

$$N = N_{100\%}\left(\frac{100}{\mathrm{DoD}}\right)L \tag{17.6}$$

where DoD is depth of discharge, N is the cycle lifetime index, and L is the lifetime curve.

17.2 Characteristics of Li-Ion Battery

Various lithium-ion batteries are lithium cobalt oxide, lithium manganese oxide, lithium nickel manganese cobalt oxide, and lithium nickel cobalt aluminium oxide and lithium iron phosphate. Depending on the temperature and various conditions, the migration of lithium ions happened during the battery charging and discharging between the cathode and anode which influences the exchange of electrons by doping and dedoping. During charging/discharging, the reaction occurs from cathode to anode and vice versa, respectively.

During discharge,

$$\mathrm{LiCoO_2 + C \rightarrow Li_{1-x}CoO_2 + Li_xC} \quad (15.7)$$

During charge,

$$\mathrm{Li_{1-x}CoO_2 + Li_xC \rightarrow Li_{1-x+dx}CoO_2 + Li_{x-dx}C} \quad (15.8)$$

17.3 Simulation Results

The simulation of battery modelling and charging/discharging characteristics of the batteries is realized in the MATLAB environment. In this paper, two case studies are performed.

Case Study 1: Study of single battery charge/discharge characteristics

Case Study 2: Study of the electric vehicle batteries of a community

Case Study 1: Study of single battery charging and discharging characteristics. The assumed specification of the battery is listed as follows.

- Rated power—960 W
- Open-circuit voltage—45 V
- Nominal voltage—48 V
- Rated capacity—20 Ah

The concept of the algorithm developed for the battery charging/discharging is shown in Fig. 17.2. When the battery's SoC is less than 10%, the battery needs to be charged; hence, it moves to charge mode. Similarly, SoC is greater than 90%, and the battery has enough charge; the battery has capable to charge other devices. It means that the DoD is taken as 0.1 and SoH is assumed as 0.9. Figure 17.3 shows the SoC of a battery with respect to time. When SoC reaches 10%, the battery is charged again. When it reaches 90%, the discharge happens.

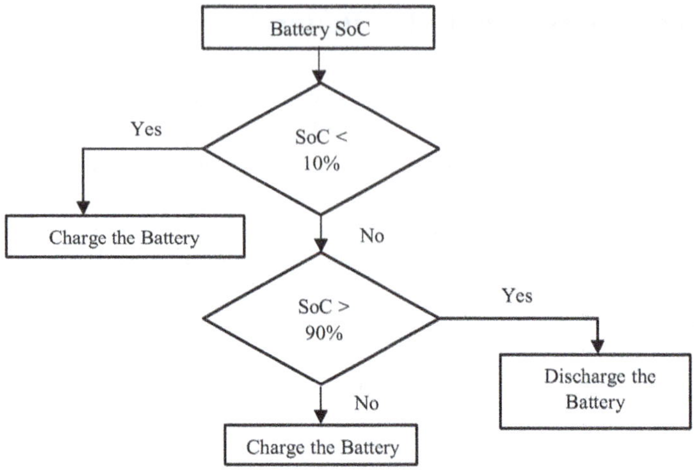

Fig. 17.2 Flow chart for battery charging/discharging algorithm

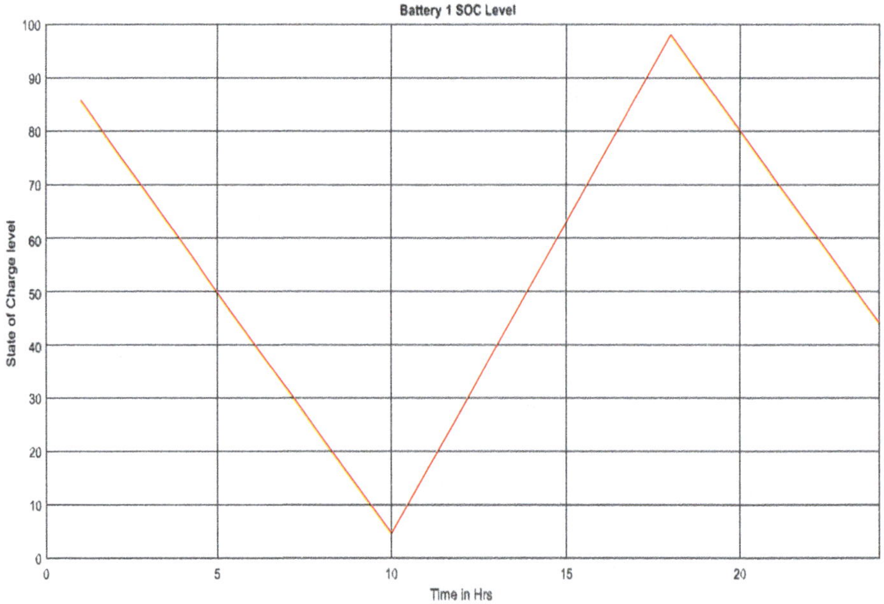

Fig. 17.3 Case study 1: time versus battery SoC charging/discharging

Figure 17.3 shows the time versus battery SoC charging and discharging characteristics curve. It is inferred that the initial SoC of a battery is 85% and hence according to the algorithm the battery started discharging up to 5% of SoC. When it reached the low SoC, the battery started charging again. The process continues based on algorithm.

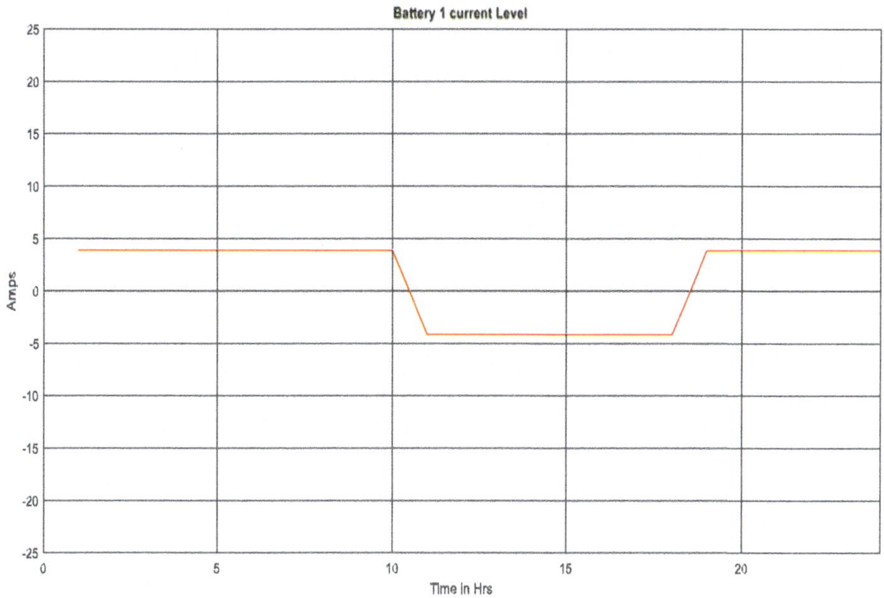

Fig. 17.4 Case study 1: time versus charging/discharging current

Figure 17.4 shows the time versus battery current during the charging and discharging conditions, in which the negative current slope indicates the battery charging and similarly positive current slope for battery discharge.

Figure 17.5 shows the battery voltage with respect to time during the charging/discharging conditions, respectively. Simultaneously, the voltage is gradually decreased that value should not go below the DoD. Hence, SoC is reached to 10%, and the discharging is stopped else which reduces the life of the battery.

Case Study 2: Study of the electric vehicle batteries' of a community

In this study, it is assumed that thirty electric vehicles are available in a community that has three apartment buildings. Building 1, Building 2, and Building 3 have 15 electric vehicles, 5 EVs, and 10 EVs, respectively. The ratings of the building 2 EVs are given in Table 17.1 which is selected based on the commercial EV. The assumption made in this paper is that, if the SoC is below 10%, the life of the battery gets reduced. Hence, minimum SoC is fixed 10%. Similarly, 90% is fixed as the upper limit of the SoC.

Figure 17.6 indicates SoC of five EV batteries. When the SoC is reached to 90%, the battery starts discharging, and it is shown in Fig. 17.6. Similarly, when it reaches 10% of SoC, it starts charging. During charging and discharging of the battery, its current and voltage waveforms are shown in Fig. 17.7 and 17.8, respectively. At the charging condition, the current in the battery is negative value that was obtained in Fig. 17.7. This indicates that the battery is considered as the load for the network. Whereas in the discharging condition, the battery acts as the source, and hence, the

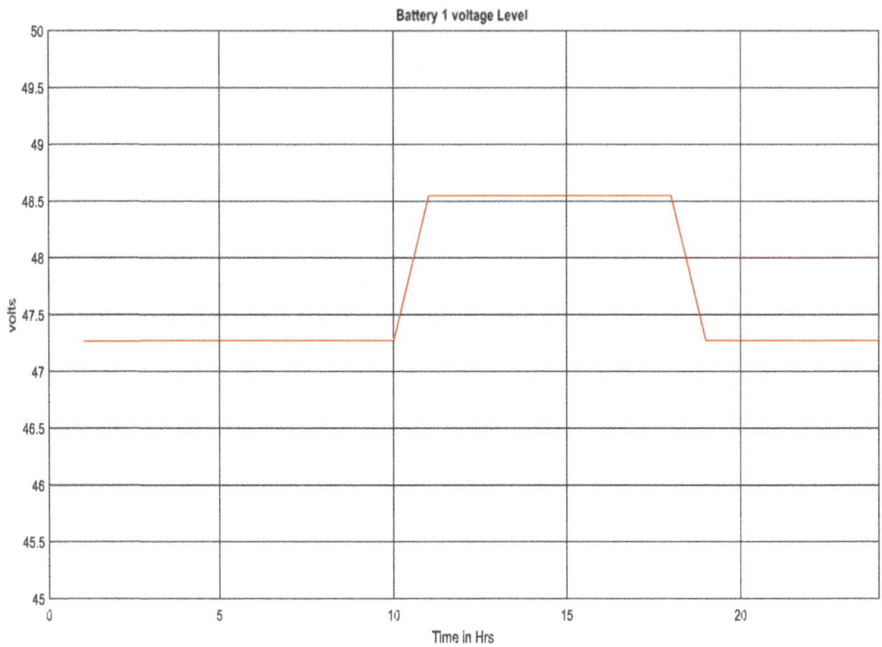

Fig. 17.5 Case study 1: time versus charging/discharging voltage

Table 17.1 Electric vehicle specification of building 2

Battery vehicle numbers	SoC (%)	Nominal voltage (V)	Battery capacity (Ah)	Open-circuit voltage (V)
1	30	48	20	45
2	50	48	20	45
3	70	48	33	45
4	90	48	20	45
5	10	48	24	45

battery current is a positive value. At the charging condition, the battery voltage value is negative, in discharging, the voltage value is positive, and it is shown in Fig. 17.8. The battery capacity is indicated by means of battery power which is the product of battery voltage and current. The battery capacity of the EVs available in the community is estimated which is shown in Fig. 17.9. The negative/positive power indicates that the battery is in charging/discharging mode, respectively. It is inferred that the available power requirements are at the instant of time. This concept is very helpful in the power system network for V2G and G2V applications.

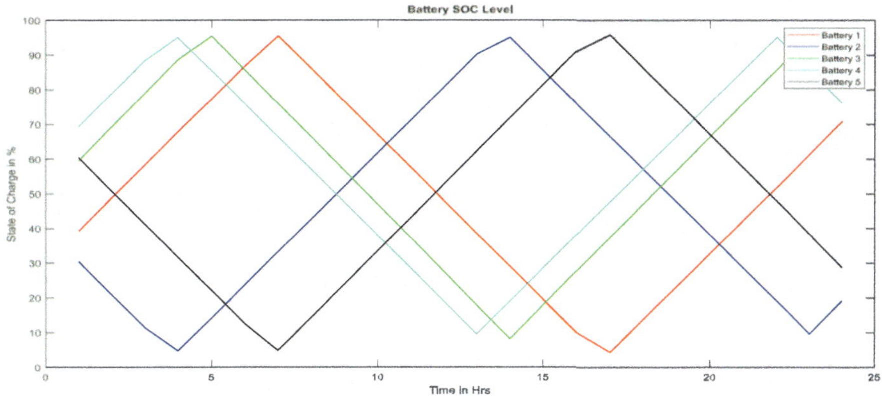

Fig. 17.6 Case study 2: time versus SoC charging/discharging

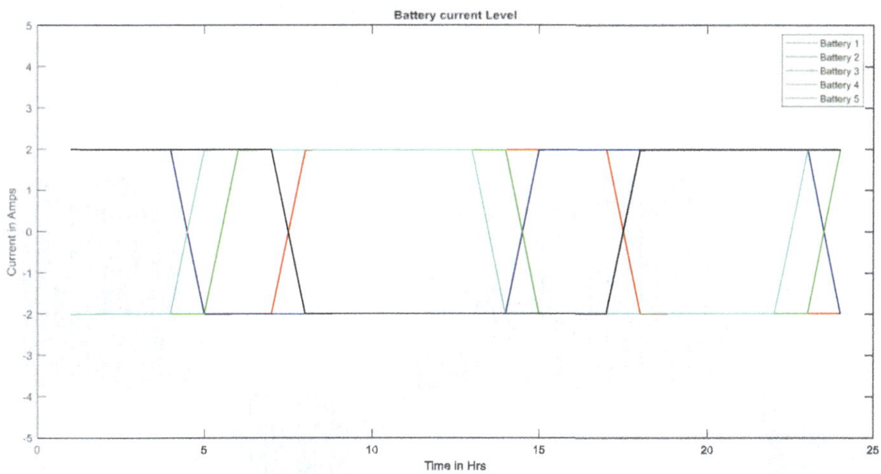

Fig. 17.7 Case study 2: time versus charging/discharging current

17.4 Conclusion

The LIB chemical reaction and its Thevenin's equivalent model are presented. Battery charging/discharging characteristics are achieved with MATLAB. The SoC, battery current, and voltage waveforms are obtained for a single battery and electric vehicle batteries available in a community. The battery current is negative/positive during the charging/discharging of the battery. The calculated battery power indicates the charging and discharging conditions. If the power is negative/positive, then the battery is discharging/charging mode which represents that it acts as a source/load, respectively. This concept is useful for the microgrid systems which consist of electric

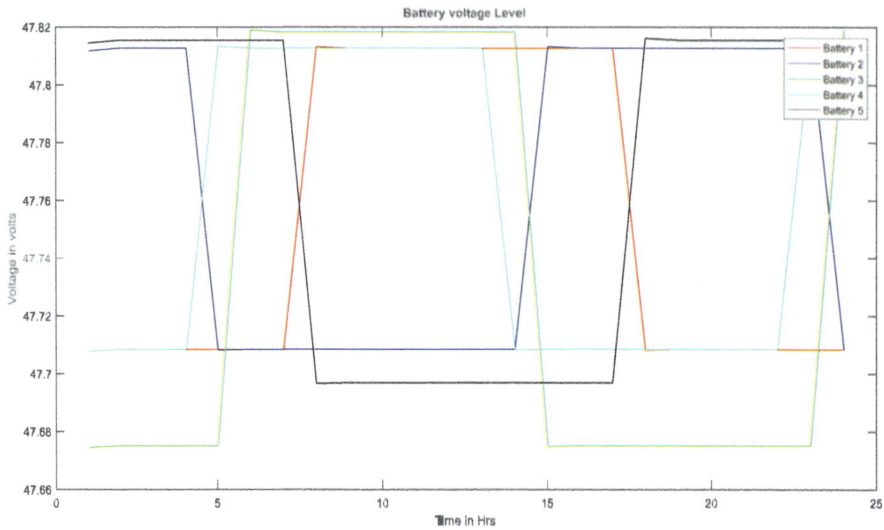

Fig. 17.8 Case study 2: time versus charging/discharging voltage

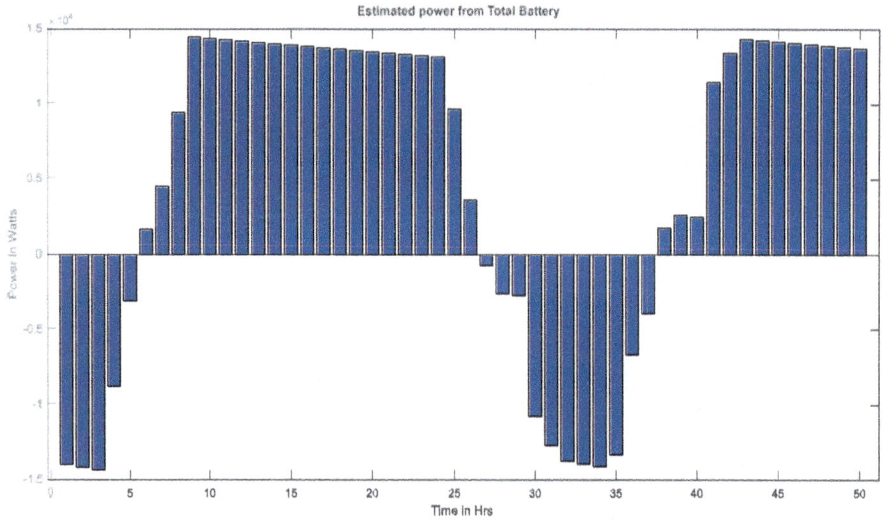

Fig. 17.9 Case study 2: time versus total power capacity of batteries in community

vehicles. The electric vehicle charging and discharging operation could be used for V2G and G2V based on the microgrid requirements.

References

1. B. Kennedy, D. Patterson, S. Camilleri, Use of lithium-ion batteries in electric vehicles. J. Power Sources **90**, 156–162 (2000)
2. G.L. Plett, Extended Kalman filtering for battery management systems of LiPB-based HEV battery packs: part 2. Modeling and identification. J. Power Sour. **134**, 262–276 (2004)
3. M. Dubarry, B.Y. Liaw, M.-S. Chen, S.-S. Chyan, K.-C. Han, W.-T. Sie, S.-H. Wu, Identifying battery aging mechanisms in large format Li-ion cells. J. Power Sour. **7**, 656–673 (2010)
4. X. Rui, H. Hongwen, W. Yi et al., Study on Ultra capacitor-battery hybrid power system for PHEV applications. High Tech. Lett. **16**, 23–28 (2010)
5. A. Saber, G. Venayagamoorthy, Efficient utilization of renewable energy sources by gridabel vehicles in cyber-physical energy systems, IEEE Syst. J. **4**(3), 258–294 (2010)
6. A. Saber, G. Venayagamoorthy, Plug in vehicles and renewable energy sources for cost and emission reductions. IEEE Trans. Ind. Electron. **58**(4), 1229–1238 (2011)
7. H. Rahimi-Eichi, U. Ojha, F. Baronti, M. Chow, Battery management system: an overview of its application in the smart grid and electric vehicles. IEEE Ind. Electron. Mag. **7**, 4–16 (2013)
8. R. Xiong, H, He, J. Fan, Online SOC estimation of high-power lithium ion batteries used on HEVs, in *Industrial Electronics Magazine, IEEE* (2011)
9. H. Hongwen, X. Rui, F. Jinxin, Evaluation of lithium-ion battery equivalent circuit models for state of charge estimation by an experimental approach. Energies **4**, 582–598 (2011)
10. F. Sun, R. Xiong, H. He, W. Li, J.E.E. Aussems, Model-based dynamic multi-parameter method for pek power estimation of li-ion batteries, *Appl. Energy* **96**(3), 378–386 (2012)
11. M.M. Hoque, M.A. Hannan, A. Mohamed, Model development of charge equalization controller for lithium-ion battery. Adv. Sci. Lett. **23**, 5255–5259 (2017)
12. H. Jan Bergveld, W.S. Kruijt, P.H.L. Notten, Battery management systems design by modelling, in *Battery Management Systems Design by Modelling* (2002), pp. 191–210
13. B.S. Bhangu, P. Bentley, D.A. Stone, C.M. Bingham. Observer techniques for estimating the state-of-charge and state-of-health of VRLABs for hybrid electric vehicles, in *IEEE Conference Vehicle Power and Propulsion* (2005), pp. 780–789
14. F. Zhang, G. Liu, Member, L. Fang, A battery state of charge estimation method with extended Kalman filter, in *Proceedings of the 2008 IEEE/ASME*, Xi'an, China, (2008), 1008–1013
15. G.L. Plett, Lib; Dynamic cell models for Kalman-filter SOC estimation, in *Proceedings of the 19th International Electric Vehicle Symposium*, Bussan, Korea (2009), pp. 193–204

Chapter 18
PMBLDC Motor Design and Analysis for Automotive Applications

R. Ruthra Prakashini, Aneesh Jategaonkar, Kareti Aasritha, and N. C. Lenin

Abstract Modeling of brushless direct current (BLDC) motor for the application of lumbar support, headrest and lift gate is carried out in this paper. Two separate models have been simulated using finite element software packages for the applications. One of the models is based on surface-mounted configuration of the BLDC motor, while the other is based on spoke-type configuration. The surface-type and spoke-type models procure their names based on the mounting of permanent magnets, which are an integral component of any BLDC machine. Extensive comparison has been carried out between the designed models. Critical parameters such as constituent weights, efficiency, output power, torque, speed, etc. are compared. Performances of the designed machines are depicted using graphical portrayal. Parameters such as locked rotor torque, no load speed, maximum efficiency, maximum output power, etc. are also compared.

Keywords Brushless direct current (BLDC) motor · Finite element analysis · Spoke type BLDC motor · Surface mounted BLDC motor

18.1 Introduction

BLDC motor is most preferred as it has the capability to provide large volume of torque for wide range of speeds and compact in size. So, it is most preferred machine for automotive application and has gained more popularity over few decades. Permanent magnet BLDC motors are being extensively used in automotives because of higher efficiency and operation flexible [1]. As the name implies, BLDC motors are electronically commutated and they do not use brushes for commutation. Compared to induction motors and brushed DC motors BLDC motors have many advantages over them. A few advantages of BLDC motors are: high-speed range, noiseless functioning, better speed versus torque characteristics, high efficiency, long running life [2–4]. Furthermore, making it productive the ratio of torque delivered to the size

R. Ruthra Prakashini · A. Jategaonkar · K. Aasritha · N. C. Lenin (✉)
Vellore Institute of Technology, Chennai Campus, Chennai 600127, India
e-mail: lenin.nc@vit.ac.in

© The Author(s), under exclusive license to Springer Nature Singapore Pte Ltd. 2021
V. Kamaraj et al. (eds.), *Emerging Solutions for e-Mobility and Smart Grids*,
Springer Proceedings in Energy, https://doi.org/10.1007/978-981-16-0719-6_18

Fig. 18.1 Lumbar support fixed in back seat [9]

of motor is higher, which helps to reduce the space and weight constraints of the motor.

18.1.1 Lumbar Support Motor

With the increase in number of driver on road comes in the increase complaints about lower back pain problems. As the time spent while driving has been increased, improper posture will cause discomfort to the driver and also increases the lower back pain. Lumbar support helps to sit up straight rather than a slumped position which benefits for proper posture [5]. It has become one of the criteria buyers look up on. BLDC motors are an appealing prospect for lumbar support for its low audible voice and overload protection. Figures 18.1 and 18.2 show the lumbar motor which is used in automobile.

18.1.2 Headrest Adjuster Motor

It is a control system method to adjust the position of headrest of a seat. A headrest has become attached to support rods. The support rods are inserted in the backrest pushed and arranged longitudinally to the backrest. In the seat notice that the height the headrest is not properly adjusted for individual needs, it is difficult to move the headrest from sitting to a new adjust. So, in some countries beginning in the late 1960s, head restraints have prevented or mitigated thousands of serious injuries

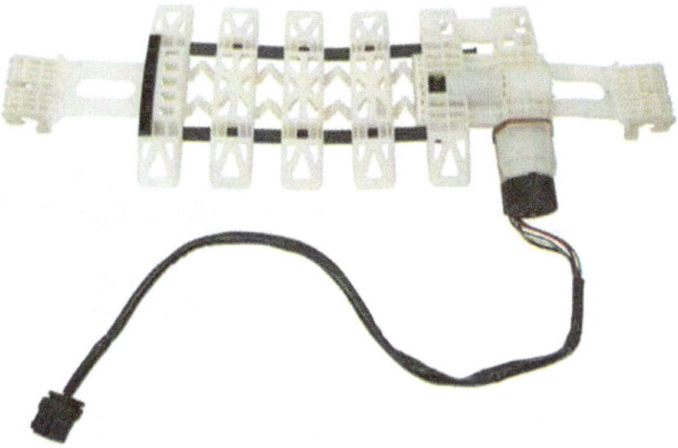

Fig. 18.2 Lumbar support motor [10]

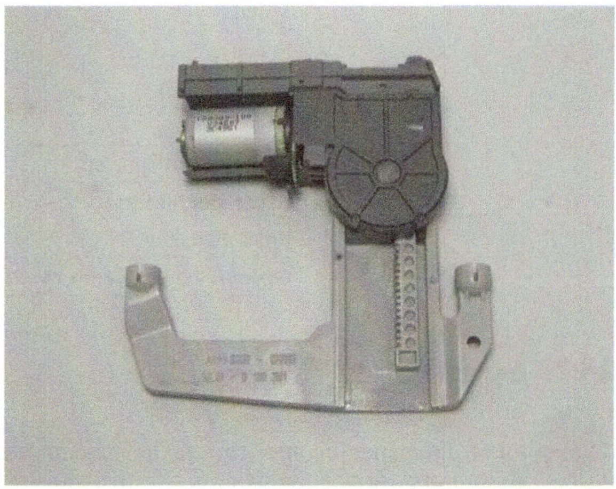

Fig. 18.3 Headrest adjuster [11]

to prevent or mitigate whiplash or injury to the cervical vertebrae [6]. The desired characteristics for BLDC motor for application in headrest are due to its high power and low noise. Figure 18.3 shows the headrest motor system used in application.

18.1.3 Power Lift Gate Motor

Power lift gate is a back-door system. The system is depicted in Fig. 18.4. The user of the vehicle controls the lift gate opening and closing by pressing a key. The keys are located either on the dashboard or handheld remote. Many power lift gate systems require the touch of a button to work, although hands-free options are becoming more available on a range of vehicles [7]. The hands-free power lift gate also offers intelligent anti-trap, height memory function and other functions.

Fig. 18.4 Power lift gate motor [12]

Table 18.1 Specifications of lumbar support motor

Quality	Values
Output power	13.64 W
Voltage	12 V
Speed	4700 rpm
Torque	0.027 N m
Stack length	35.8 mm
Stator outer diameter	65 mm
Stator slot fill factor	45%
Shaft diameter	3.175 mm
Operating Temperature	−40 to 85 °C

18.2 Specification

The motors for the specific applications have been designed for both spoke-type and surface-mounted lumbar support motor with same specifications. Table 18.1 depicts the specifications for lumbar motor which satisfies the constraints of low audible voice and overload protection. BLDC motor is designed with the property of low noise and high powers for headrest adjustment, and its specifications are shown in Table 18.2. Power lift gate specifications are depicted in Table 18.3.

18.3 Design Approach

The BLDC motor for spoke type and surface mounted for the applications lumbar support, headrest adjuster and power lift gate has been designed with steel type of stator and rotor core as M36_29G. Ceramic8D is used as its permanent magnet, and for winding, copper is being used as its material.

Table 18.2 Specifications of headrest adjustment motor

Quality	Values
Output power	5.57 W
Voltage	12 V
Speed	3208 rpm
Torque	0.01669 N m
Stack length	35.8 mm
Stator outer diameter	50 mm
Stator slot fill factor	45%
Shaft diameter	3.175 mm
Operating temperature	−40 to 85 °C

Table 18.3 Specifications of power lift gate motor

Quality	Values
Output Power	46.80 W
Voltage	12 V
Speed	8147 rpm
Torque	54.86 N m
Stack Length	87.3 mm
Stator outer diameter	35.9 mm
Stator slot fill factor	45%
Shaft diameter	3.175 mm
Operating temperature	−40 to 85 °C

18.3.1 Stator Design

The number of slots used for both spoke type and surface mounted varies accordingly to the number of poles in order to achieve the design specification. The slot types used for all the models are same, and it is depicted in Fig. 18.5. "BsO" is kept twice the diameter of the wire as it would be easier for winding. "Bs1 and "Bs2" help to adjust the width of the slots to make slots the adjacent slot to approximately parallel with it. "BsO" and "HsO" size variations play an important role in the core losses. "Hs2" varies the depth the slot. For practical convenience, the slot fill factor is to 45%. The slot configuration for the application lumbar support, headrest adjuster and power lift gate is depicted in Tables 18.4, 18.5 and 18.6, respectively.

Fig. 18.5 Stator slot model

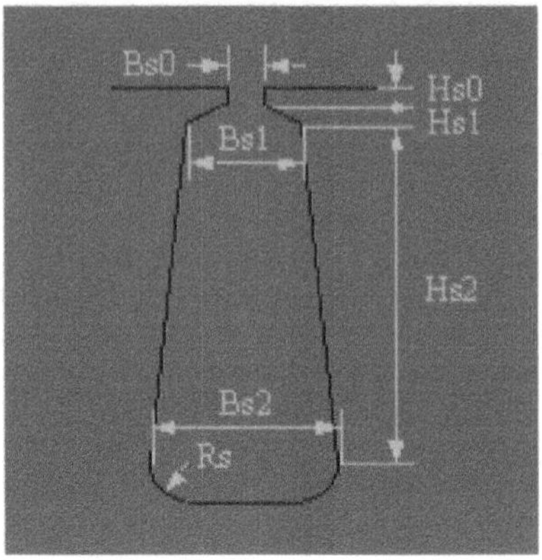

Table 18.4 Slot configuration of lumbar support motor

Parameters	Surface mounted	Spoke type
Hs0	0.2	0.6
Hs1	0.3	0.3
Hs2	3.5	4.3
Bs0	1.2	1
Bs1	3.5	3.7
Bs2	5	5.5
Rs	1	2

Table 18.5 Slot configuration of headrest adjustment motor

Parameters	Surface mounted	Spoke type
Hs0	0.2	0.3
Hs1	0.2	0.2
Hs2	4	4.9
Bs0	1	1.15
Bs1	3	4.1
Bs2	5	7
Rs	1	2.5

Table 18.6 Slot configuration of power lift gate motor

Parameters	Surface mounted	Spoke type
Hs0	0.2	0.1
Hs1	0.2	0.1
Hs2	3.5	2.5
Bs0	1.6	1.29
Bs1	3	7.8
Bs2	5	10.5
Rs	1	1.8

18.3.2 Rotor Design

Permanent magnet is mounted on the outer surface of the rotor in two arrangements. Those are surface-mounted and spoke-type model. The major weight and cost of the motor are contributed by permanent magnet. Surface-mounted and spoke-type model are depicted in Figs. 18.6 and 18.7, respectively. Material used for permanent magnet is Ceramic8D; it is insusceptible to demagnetization by outer fields. Ceramic8D is stronger and inexpensive than few natural magnets, thus reduces the overall cost of the magnet. The models are designed with minimum of 0.5 mm gap between the outer diameter of rotor and inner diameter of the stator. The magnets are inserted on the rotating shaft, which in turn maximizes the output traits of the air gap flux density

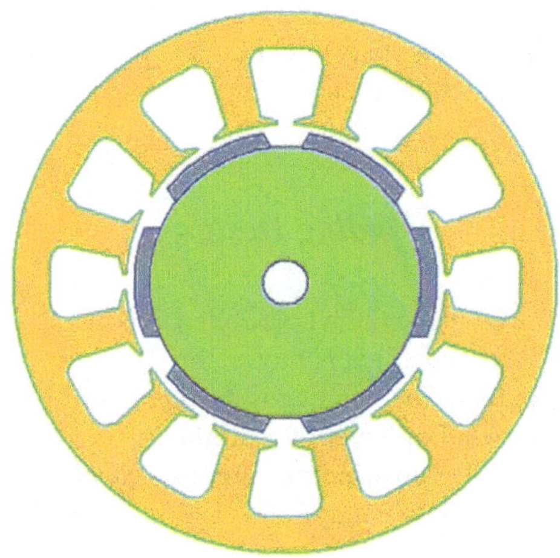

Fig. 18.6 Lumbar support surface-mounted model

Fig. 18.7 Lumbar support spoke-type model

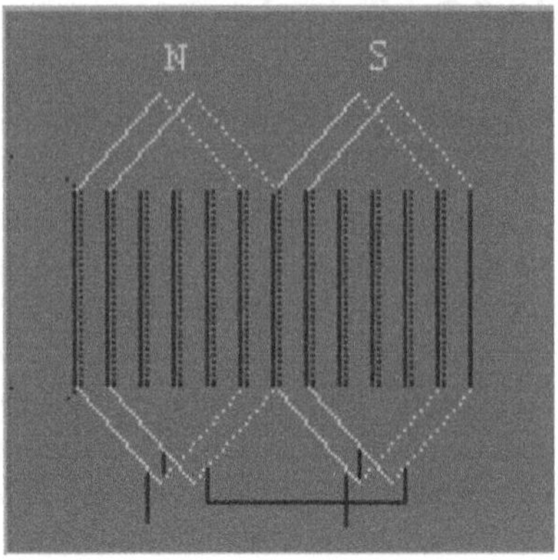

Fig. 18.8 Lap winding

[7]. The magnetic thickness and the magnetic weight have been varied to achieve the output requirements.

18.3.3 Winding Design

The efficiency of whole coiled winding is greater than half-coiled wound motors [8]. To have an efficient motor, the BLDC motors have been designed with whole coiled winding. The winding is depicted in Fig. 18.8. The motors have been designed with two winding layers for all the applications. Lap winding has been used for all the designs. Lap windings are also called as parallel winding. Lap winding has been used for low voltage and also increases the current carrying range.

18.4 Results

Parameters such as output power, input power, efficiency, locked rotor torque, no load speed, maximum efficiency, maximum output power, etc. are compared. The weights of materials used in the rotor and stator like permanent magnet weight, armature copper weight, etc. are also compared to view advantages of motor for different applications.

18.4.1 Lumbar Support

The output results of surface-mounted and spoke-type BLDC lumbar support are shown in Table 18.7; Figs. 18.9, 18.10 and 18.11. The maximum torque in spoke-type model is 0.174 N m which is relatively higher than vividly used lumbar support

Table 18.7 Performance comparison—spoke-type and surface-mounted lumbar support motor

Parameters	Surface mounted	Spoke type
Average input current (A)	1.7009	1.84995
Total loss (W)	6.64658	8.20726
Output power (W)	13.7642	13.9921
Input power (W)	20.4108	22.1994
Efficiency (%)	67.436	63.0293
Rated speed (rpm)	4700	4700
Rated torque (N m)	0.0279657	0.028429
Permanent magnet weight (gm)	12.9651	8.51032
Armature copper weight (gm)	43.5015	76.1842
Armature core steel weight (gm)	181.2	184.347
Rotor core steel weight (gm)	107.509	51.5869
Total net weight (gm)	345.175	320.628

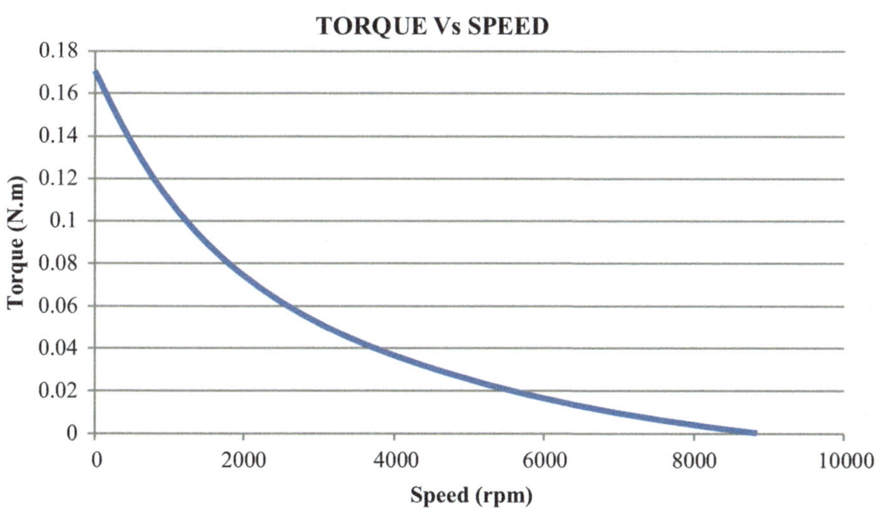

Fig. 18.9 Torque versus speed—lumbar support spoke-type model

motors. In spoke-type model, the maximum efficiency of 75% is obtained at the speed of 6800 rpm. In Figs. 18.12, 18.13 and 18.14, comparison of weight between surface-mounted and spoke-type model is depicted. The permanent magnet weight and overall weight of spoke-type model is comparatively lesser than surface-mounted model. The efficiency obtained by surface-mounted model is 7.4176% greater than spoke-type model.

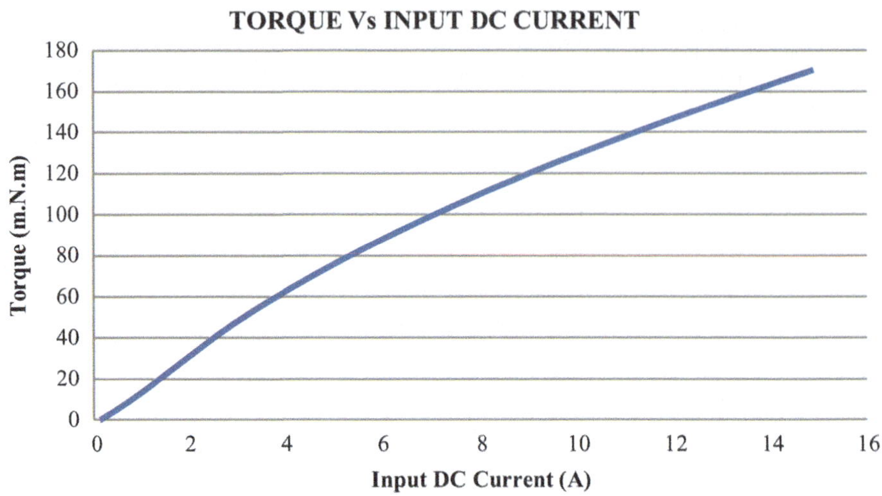

Fig. 18.10 Torque versus input DC current—lumbar support spoke-type model

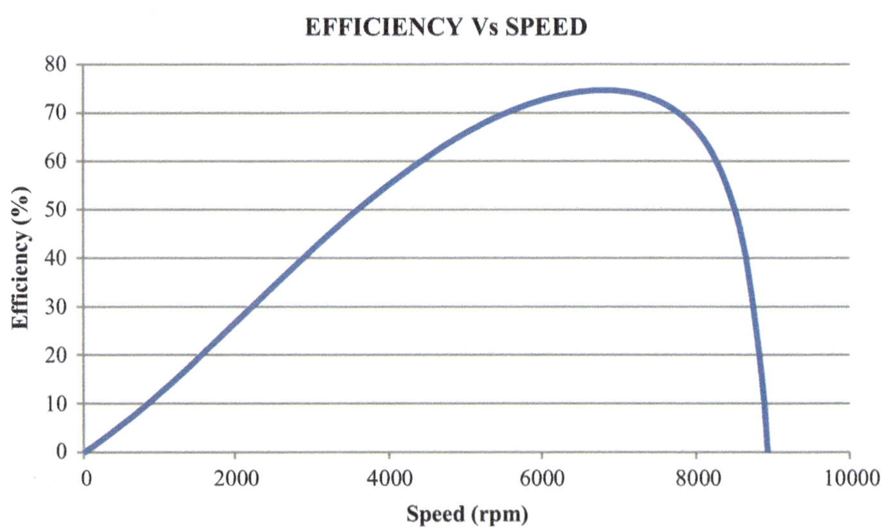

Fig. 18.11 Efficiency versus speed—lumbar support spoke-type model

18.4.2 Headrest Adjuster

The results of headrest adjuster motor are shown in Table 18.8; Figs. 18.15, 18.16, 18.17, 18.18, 18.19, 18.20 and 18.21. In Fig. 18.15, the torque takes up a parabolic gradual decline as the speed increases. The critical point, i.e., 70% efficiency, is

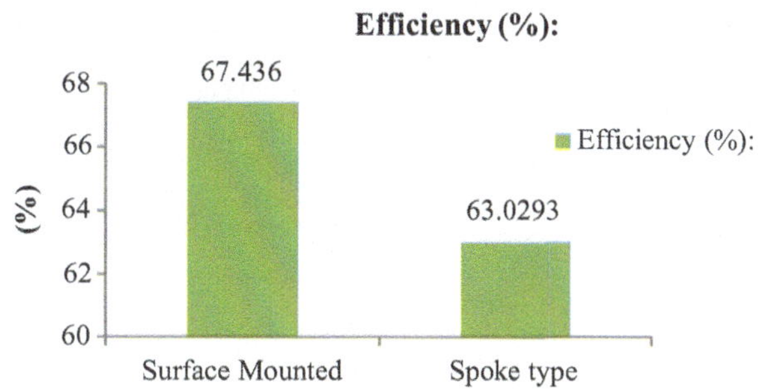

Fig. 18.12 Efficiency comparison—lumbar support spoke-type and surface-mounted model

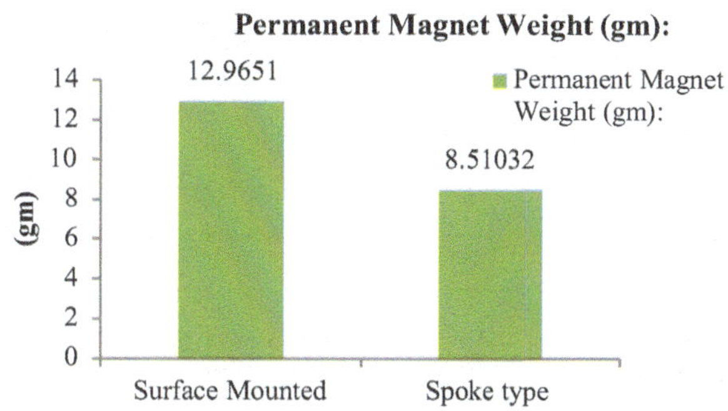

Fig. 18.13 Permanent magnet weight comparison—lumbar support spoke-type and surface-mounted model

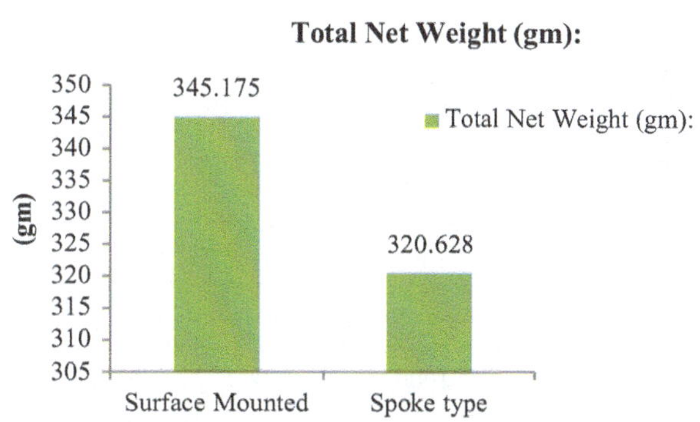

Fig. 18.14 Total net weight comparison—lumbar support spoke-type and surface-mounted model

obtained at the rpm of 4000. The spoke-type model has consumed 3.40826 g permanent magnet lesser than surface mounted, in turn spoke-type model has resulted 17.255 g lesser. The weight achieved in comparably lesser than the motors which are widely used. This helps to cut down the cost of the motor slightly. In Fig. 18.21, the flux distribution of the motor is shown.

Table 18.8 Performance comparison—spoke-type and surface-mounted headrest adjuster motor

Parameters	Surface mounted	Spoke type
Average input current (A)	0.656287	0.754401
Total loss (W)	2.20148	3.2021
Output power (W)	5.67397	5.85071
Input power (W)	7.87544	9.05281
Efficiency (%)	72.0463	64.6287
Rated speed (rpm)	3208	3208
Rated torque (N m)	0.0168898	0.0174159
Permanent magnet weight (gm)	10.2369	6.82864
Armature copper weight (gm)	37.6109	66.2452
Armature core steel weight (gm)	147.598	163.23
Rotor core steel weight (gm)	89.5907	31.4776
Total net weight (gm)	285.037	267.782

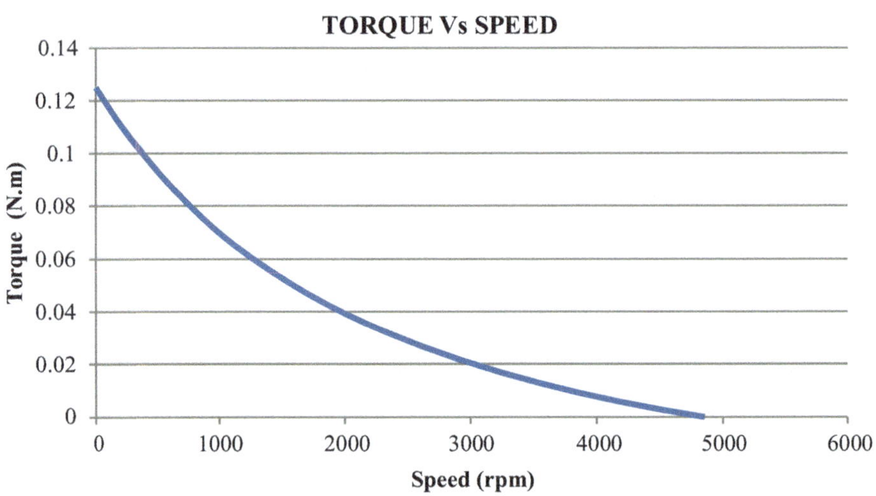

Fig. 18.15 Torque versus speed—headrest adjuster spoke-type model

18.4.3 Power Lift Gate

Table 18.9; Figs. 18.22, 18.23, 18.24, 18.25, 18.26 and 18.27 depict the output results of power lift gate design. In Fig. 18.22, there is a steep fall of torque from 0.55 to 0.1 N m, and then the torque reduces gradually. The maximum efficiency is obtained at the speed 11,260 rpm. The spoke-type model has consumed less permanent magnet

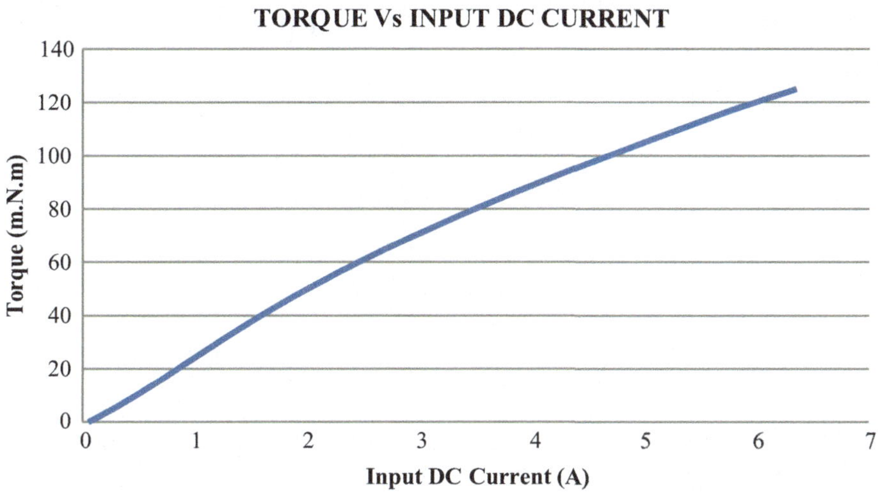

Fig. 18.16 Torque versus input DC current—headrest adjuster spoke-type model

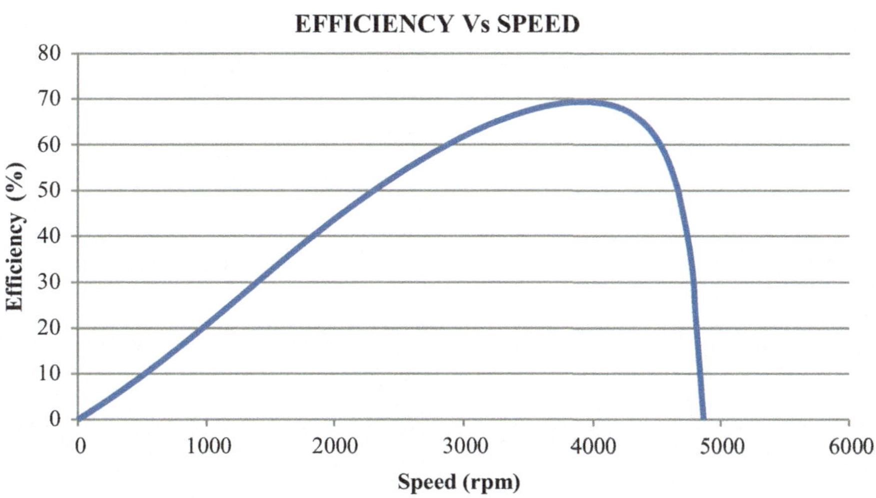

Fig. 18.17 Efficiency versus speed—headrest adjuster spoke-type model

weight with the difference of 11.7531 g, which can effectively reduce the cost of the motor. The efficiency and overall weight of surface-mounted motor are higher.

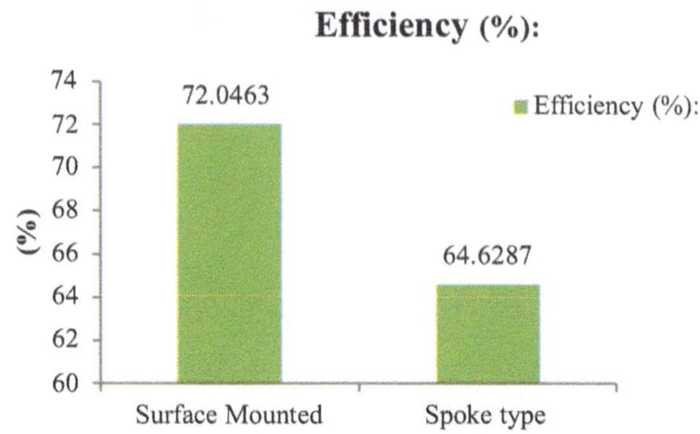

Fig. 18.18 Efficiency comparison—headrest adjuster spoke-type and surface-mounted model

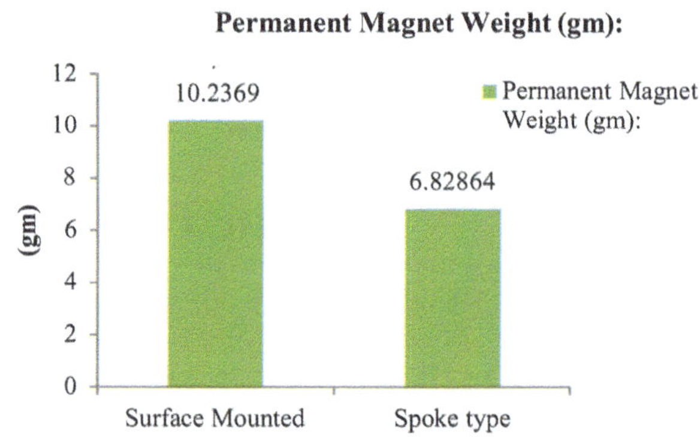

Fig. 18.19 Permanent magnet weight comparison—headrest adjuster spoke-type and surface-mounted model

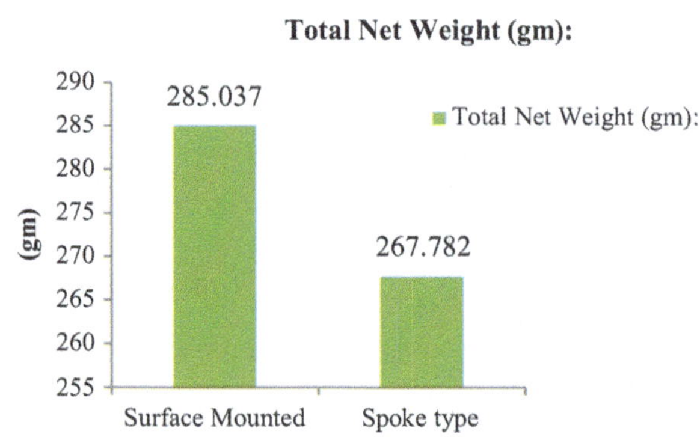

Fig. 18.20 Total net weight comparison—headrest adjuster spoke-type and surface-mounted model

18.5 Conclusion

In this paper, spoke-type and surface-mounted BLDC motor are designed for lumbar support, headrest adjuster and power lift gate application. We can conclude that the

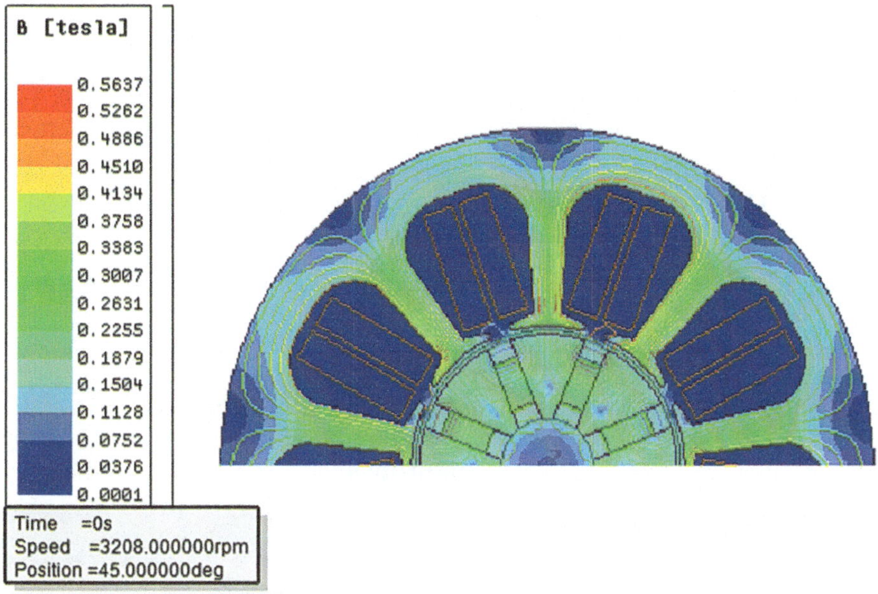

Fig. 18.21 Flux distribution of headrest adjuster spoke-type model

Table 18.9 Performance comparison—spoke-type and surface-mounted power lift gate motor

Parameters	Surface mounted	Spoke type
Average input current (A)	5.3869	6.02014
Total loss (W)	16.2106	25.9777
Output power (W)	48.4322	46.264
Input power (W)	64.6428	72.2417
Efficiency (%)	74.9228	64.0406
Rated speed (rpm)	8147	8147
Rated torque (N m)	0.05677	0.0542272
Permanent magnet weight (gm)	23.4911	11.738
Armature copper weight (gm)	40.485	67.5477
Armature core steel weight (gm)	188.322	327.834
Rotor core steel weight (gm)	69.9239	112.071
Total net weight (gm)	322.222	519.191

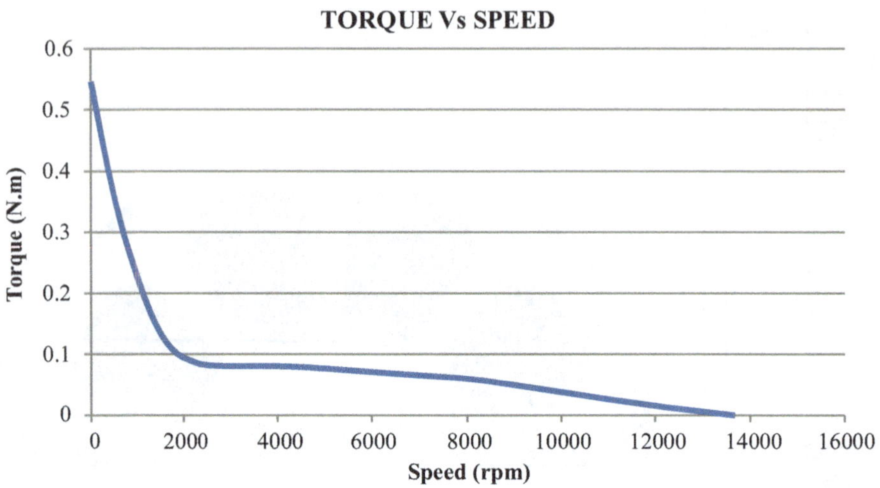

Fig. 18.22 Torque versus speed—power lift gate spoke-type model

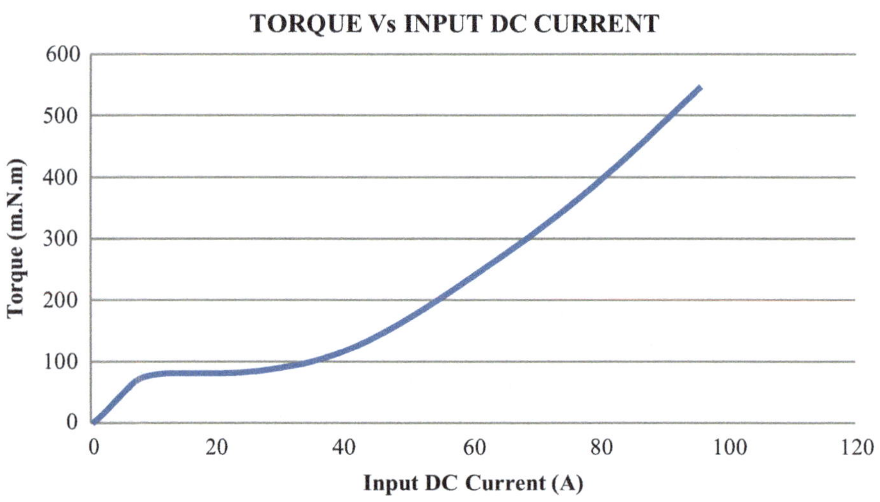

Fig. 18.23 Torque versus input DC current—power lift gate spoke-type model

overall permanent magnet weight of the spoke-type model is lesser than surface-mounted model. Ceramic 8D has been used as permanent magnet material in order to reduce the cost of permanent magnet, the major contributor of the overall cost of the motor. The average input current of spoke-type models is greater than surface-mounted model, and the efficiency of surface-mounted model is greater than spoke-type model for all the applications.

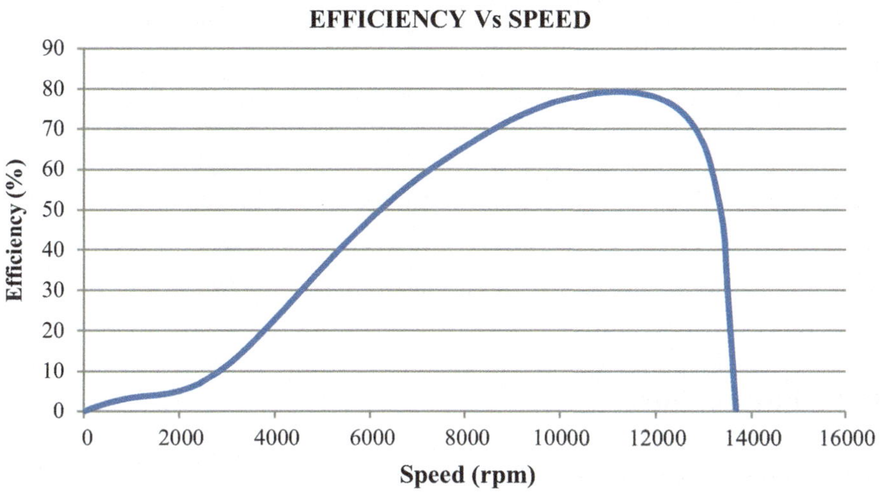

Fig. 18.24 Efficiency versus speed—power lift gate spoke-type model

Fig. 18.25 Efficiency comparison—power lift gate spoke-type and surface-mounted model

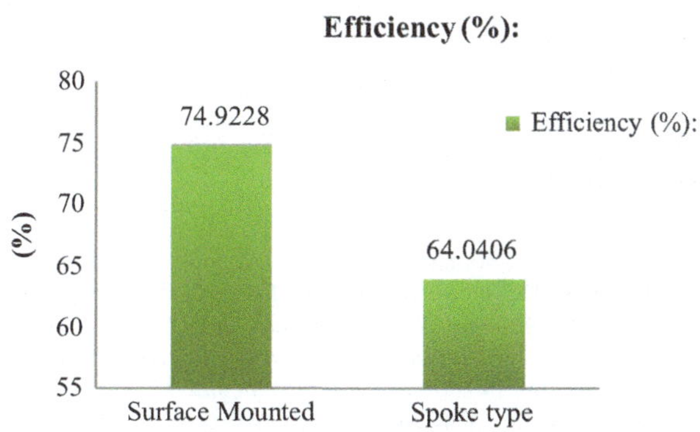

Fig. 18.26 Permanent magnet weight comparison—power lift gate spoke-type and surface-mounted model

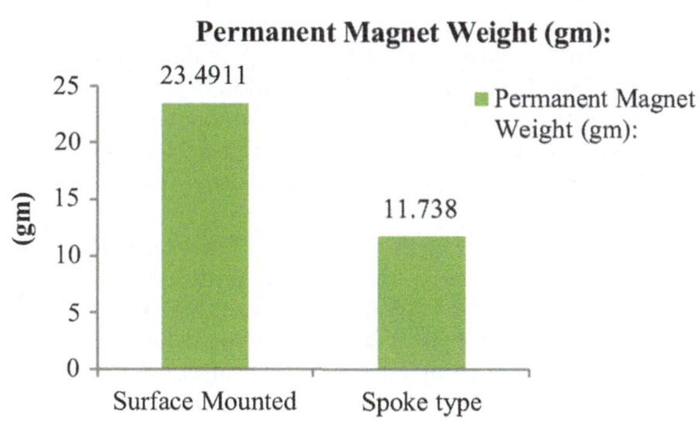

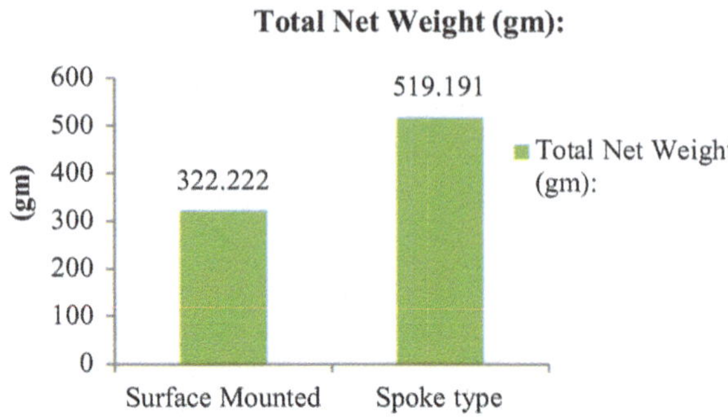

Fig. 18.27 Total net weight comparison—power lift gate spoke-type and surface-mounted model

18.6 Future Works

Future works can be carried out on designing of BLDC motor on other applications like air pump, window lift drive, anti-lock braking system (ABS), seatbelt pre-tensioner, electric parking brake, etc. The designs can be expanded by using different permanent magnet types. Future works can be done on detailed thermal characterises of motor.

References

1. D. Uygun, S. Solmaz, Design and dynamic study of a 6 kW external rotor permanent magnet brushless DC motor for electric drivetrains, in *2015 IEEE 5th International Conference on Power Engineering, Energy and Electrical Drives* (2015)
2. L. Chang, J. Muszynski, Design of a 5-phase permanent magnet brushless DC motor for automobiles, in *2003 IEEE 58th Vehicular Technology Conference* (2003)
3. A. Jategaonkar, P. Ramesh, P. Kochgabay, N.C. Lenin, Electromagnetic and thermal analyst is of permanent magnet BLDC wiper motor, in *The Advances in Electrical Control and Signal System, AECSS* (2019)
4. P. Shetty, A.K.N. Subramonium, M.S. Kumar, Mathematical modeling of permanent magnet brushless dc motor for electric scooter, in *2015 Fifth International Conference on Communication Systems and Network Technologies*, Gwalior, pp. 1222–1226 (2015)
5. H. Hasan, H. Ismail, R. Mohamed Firhad, R. Azidin, Preventive methods of low back pain, in *2010. International Conference on Science and Social Research (CSSR 2010)*, Kuala Lumpur, Malaysia (2010), pp. 1278–1282
6. J. Murgoitio, M. Ferros, A. Goti, M. Larburu, T. Rodríguez, Robotiker-Tecnalia, Networking sensors and actuators for a new active headrest, in *Advanced Microsystems f or Automotive Applications*, ed. by J. Valldorf, W. Gessner (VDI-Buch, Springer, Berlin, 2007), pp 3–46
7. A. Müller, K.P. Jankowski, Tailored modeling of a vehicle power liftgate avoiding explicit definition of body-fixed reference frames, in *Volume 6: 11th International Conference on Multibody Systems, Nonlinear Dynamics, and Control* (2015)
8. H.-W. Kim, K.-T. Kim, Y.-S. Jo, J. Hur, Optimization methods of torque density for developing the neodymium free SPOKE-type BLDC motor. IEEE Trans. Magn. **49**(5), 2173–2176 (2013)
9. https://www.toyota-4runner.org/attachments/5th-gen-t4rs/259160d1499577377-lumbar-support-motor-fix-img_3442-jpg?s=42b7d44d8ebf119b292d35969314bca2
10. https://images11.palcdn.com/hlr-system/WebPhotos/87/873/8735/8735771.jpg

11. https://cdn11.bigcommerce.com/s-hkk7s/images/stencil/1600x1600/products/15161/162826/inv_0132343173.1523301139.jpg?c=2&imbypass=on&imbypass=on
12. http://www.microplanetarygearbox.com/photo/pl19269387-professional_high_torque_gear_motor_for_power_liftgate_cables_ts16949_listed.jpg

Author Index

© The Editor(s) (if applicable) and The Author(s), under exclusive license to Springer Nature Singapore Pte Ltd. 2021

V. Kamaraj et al. (eds.), *Emerging Solutions for e-Mobility and Smart Grids*, Springer Proceedings in Energy, https://doi.org/10.1007/978-981-16-0719-6

The manufacturer's authorised representative in the EU is Springer
Nature Customer Service Centre GmbH, Europaplatz 3, 69115 Heidelberg,
Germany. If you have any concerns regarding our products, please
contact ProductSafety@springernature.com

Printed and bound by CPI Group (UK) Ltd, Croydon, CR0 4YY
30/09/2026
02225294-0007